SpringerBriefs in Computer Science

SpringerBriefs present concise summaries of cutting-edge research and practical applications across a wide spectrum of fields. Featuring compact volumes of 50 to 125 pages, the series covers a range of content from professional to academic.

Typical topics might include:

- A timely report of state-of-the art analytical techniques
- A bridge between new research results, as published in journal articles, and a contextual literature review
- A snapshot of a hot or emerging topic
- An in-depth case study or clinical example
- A presentation of core concepts that students must understand in order to make independent contributions

Briefs allow authors to present their ideas and readers to absorb them with minimal time investment. Briefs will be published as part of Springer's eBook collection, with millions of users worldwide. In addition, Briefs will be available for individual print and electronic purchase. Briefs are characterized by fast, global electronic dissemination, standard publishing contracts, easy-to-use manuscript preparation and formatting guidelines, and expedited production schedules. We aim for publication 8–12 weeks after acceptance. Both solicited and unsolicited manuscripts are considered for publication in this series.

**Indexing: This series is indexed in Scopus, Ei-Compendex, and zbMATH **

Long Xu • Yihua Yan • Xin Huang

Deep Learning in Solar Astronomy

 Springer

Long Xu
NSSC
Chinese Academy of Sciences
Beijing, China

Yihua Yan
NSSC
Chinese Academy of Sciences
Beijing, China

Xin Huang
NSSC
Chinese Academy of Sciences
Beijing, China

ISSN 2191-5768   ISSN 2191-5776 (electronic)
SpringerBriefs in Computer Science
ISBN 978-981-19-2745-4   ISBN 978-981-19-2746-1 (eBook)
https://doi.org/10.1007/978-981-19-2746-1

This Springer imprint is published by the registered company Springer Nature Singapore Pte Ltd.
The registered company address is: 152 Beach Road, #21-01/04 Gateway East, Singapore 189721, Singapore

Preface

Deep learning has been extensively developed in the last decade. It is regarded as the most promising approach for artificial intelligence (AI), triggering a revolution of entire field of computer science. The most successful examples include image classification, object detection, recognition, tracking, segmentation, natural language processing (NLP), and machine translation. Recently, we have witnessed the fast development of deep learning in astronomy, which is mostly concerned with big data processing, including filtering and archiving massive observation data, detecting/recognizing objects/events, forecasting astronomical events, imaging observational data, and image restoration.

Deep learning directly learns compact/compressed features from massive raw data, without need of prior knowledge, which is superior to handcrafted features in case prior knowledge is unavailable or unreliable. This is very common in astronomy, where most of the physical processes are unknown. It should be pointed that feature extraction from raw data is indispensable in machine learning; since raw data is tremendously high-dimensional, it is impossible to establish classification/regression model over such high-dimensional data. In addition, deep learning model is an end-to-end system without need of human intervention, which endows us with an affordable human labor consumption for manipulating massive data. Besides, deep learning could simultaneously mine implicit data knowledge from "on-site" data and exploit prior knowledge already verified through statistics, by using pre-trained model and adding constraints in loss function. Nowadays, modelling over massive data has been granted a fashionable name, data-driven. Deep learning has been repeatedly verified to be the most efficient technique for data-driven modelling. The classical deep learning models, such as AlexNet, VGG, ResNet and DenseNet, have already been well pre-trained on large-scale databases. For a specific application, a light-weight network and a few samples are sufficient for modelling. Recently, we have seen a popular usage of deep learning for exploiting both prior knowledge and data-driven, where a light-weight network was embedded into a backbone network to achieve higher efficiency than either of them separately used. Data-driven implies mining data knowledge underneath massive data. A single network cannot use the prior knowledge efficiently, while mining data knowledge simultaneously. Thus, disentangling the two tasks can make independent optimization of them easier, by employing two sub-networks each of which is responsible for each task. For example, a lightweight subnetwork was embedded into a backbone network to realize super-resolution with both high-fidelity image content and elaborate gradients/edges.

Inspired by success of deep learning in computer vision, we investigated the applications of deep learning in solar big data processing: (1) object detection, e.g., sunspot, filament, active region, and coronal hole detection; (2) time series analysis, e.g., solar flare, solar radio flux, and CME forecasts; and (3) image generation, e.g., image deconvolution, deblur, desaturation, super-resolution, and magnetogram generation.

In this book, a comprehensive overview of the deep learning applications in solar astronomy, concerning concept and details of network design, experimental results, and competition with state of the art are reported and discussed for the first time. The content of this book comes from our previous publications and the latest progress on this topic. This book is expected to be a good reference for students and young researchers who are majoring in astronomy and computer science, especially the ones who are conducting interdisciplinary research.

The content of this book is arranged as follows. The first chapter gives a brief introduction of AI, deep learning, and big data of solar astronomy. The second chapter presents the classical deep learning models in the literature. Then, deep learning applications in solar astronomy are classified into four categories: classification, object detection, image generation, and forecasting, each of which is presented in a separate chapter, from the third chapter to the sixth chapter.

Beijing, China   Long Xu
Beijing, China   Yihua Yan
Beijing, China   Xin Huang
March 2022

Acknowledgments

This work was supported by the National Natural Science Foundation of China (NSFC) under Grants 11790301, 11790305, 11873060 and U1831121, the National Key R&D Program of China under Grant 2021YFA1600504, the Beijing Municipal Natural Science Foundation under Grant 1202022.

Contents

About the Authors

Long Xu received his PhD degree from the Institute of Computing Technology, Chinese Academy of Sciences (CAS), in 2009. He was selected into the 100-Talents Plan of CAS in 2014. From 2014 to 2022, he was with the National Astronomical Observatories, CAS. He is currently with both National Space Science Center, CAS and Peng Cheng Laboratory. His research interests include image/video processing, solar radio astronomy, wavelet, machine learning, and computer vision. He has published more than 100 academic papers, and also a book, titled Visual Quality Assessment by Machine Learning, with Springer in 2015.

Yihua Yan received his PhD degree from the Dalian University of Technology in 1990. He was a foreign research fellow with the NAOJ (Japan) from 1995 to 1996 and an Alexander von Humboldt Fellow with the Astronomical Institute, Wurzburg University, Germany, from 1996 to 1997. He was the President of IAU Division E: Sun and Heliosphere from 2015 to 2018. He was the director of the CAS Key Laboratory of Solar Activity (2008–2019) and the director of Solar Physics Division (2013–2021) at NAOC. He is currently a professor and chief scientist at the National Space Science Center, Chinese Academy of Sciences.

Xin Huang received his PhD degree from Harbin Institute of Technology in 2010. He was an associate professor at Solar Activity Prediction Center, NAOC, in 2013. Now, he is with the Space Environment Prediction Center, National Space Science Center, Chinese Academy of Sciences. His research interests include data mining, image processing, and short-term solar activity forecasting. He has published more than 20 academic papers, including one of the top 1% most cited papers in IOP Publishing's astrophysics journals, published over the period of 2018-2020.

Acronyms

AI	Artificial intelligence
SVM	Support vector machine
MAP	Maximum a posterior estimation
MLE	Maximum likelihood estimation
GCN	Graph convolutional network
FCN	Fully connected neural network
DBN	Deep belief network
RBM	Restricted Boltzmann machines
PCA	Principal component analysis
GRU	Gated recurrent unit
SGD	Stochastic gradient descent
BP	Back propagation
CNN	Convolutional neural network
GAN	Generative adversarial network
RNN	Recurrent neural network
LSTM	Long short-term memory
AE	Auto-encoder
R-CNN	Region-based convolutional neural network
YOLO	You only look once
RPN	region proposal network
CV	Computer vision
NLP	Natural language processing
HVS	Human vision perception system
DPM	Deformable part-based model
HOG	Histogram of oriented gradients
EUV	Extreme ultra-violent
AR	Active region
CME	Coronal mass ejection
PSNR	Peak signal to noise ratio
SSIM	Structural similarity index
SDO	Solar Dynamics Observatory

HMI	Helioseismic and Magnetic Imager
AIA	Atmospheric Imaging Assembly
SOHO	Solar and Heliospheric Observatory
MDI	Michelson Doppler Imager
EIT	Extreme ultraviolet Imaging Telescope
STEREO	Solar TErrestrial RElations Observatory
LASCO	Large Angle and Spectrometric COronagraph
SECCHI	Sun Earth Connection Coronal and Heliospheric Investigation
SKA	Square Kilometre Array
SBRS	Solar Broadband Radio Spectrometer
MUSER	MingantU SpEctral Radioheliograph
FAST	Five-hundred-meter Aperture Spherical radio Telescope
NASA	National Aeronautics and Space Administration
ESA	European Space Agency
NOAA	National Oceanic and Atmospheric Administration
SWPC	Space Weather Prediction Center of National Oceanic
JSOC	Joint Science Operations Center

Chapter 1
Introduction

Abstract Deep leaning has been developing very fast in recent years due to big data, high-performance computing and the breakthrough of neural network training techniques. It has been particularly successful in computer vision, machine translation, speech recognition and natural language processing. Modern astronomy concerns a big data challenge owning to high-resolution, high-precision and high-cadence telescopes. The big data presents a great challenge to data processing, statistical analysis and scientific discovery. Therefore, it is highly demanded to develop artificial intelligent algorithms to process big data aromatically, further discover complex relationship and mine knowledge hidden in massive data. As the best representative of artificial intelligence, a bunch of classical models have been developed for processing single image, video, speech and natural language. Among of them, convolutional neural network has been verified most efficient for processing image. To process time series input, like video, recurrent neural network, e.g., long short-term memory (LSTM), was developed, which was widely known for forecasting the future, e.g., event occurrence, physical parameter prediction. An overview of artificial intelligence, deep learning and astronomical big data is presented in this chapter, as the background of this book.

Keywords Artificial intelligence (AI) · Deep learning · Solar astronomy · Big data

1.1 Artificial Intelligence (AI)

Artificial intelligence (AI) has gained extensive attentions in various field recently, especially, computer vision (CV), speech recognition, machine translation and natural language processing (NLP). It has been widely accepted that there were three waves of AI development in history. Now, we are in the third AI wave, which started from 2006 along with the rise of deep learning. Deep learning was regarded as the most powerful and promising tool for solving AI tasks in nowadays and near future. Its boom was driven by the rise of big data, high performance computing, especially GPU computing. The first and second waves of AI occurred in 1950s and 1980s. The representative techniques were perceptron and back propagation (BP), respectively. Deep learning inherits from the well-known neural network. Previously, limited by computer capability and data resource, neural network only had several layers, and even replaced by a lightweight shallow model, namely support vector machine (SVM) in 1980s and 1990s. With the fast rise of big data, computer hardware upgrade (especially, graphics processing unit (GPU)), deep neural network (deep learning), got widespread attention from both academic and industrial fields. Big data caters for sufficient data source for feeding deep model; GPU computing contributes high computing capability for training deep model. From empirical evidence, deeper network has greater ability of nonlinear fitting for mining high-level semantic information of input data, contributing a compact representation of input data, which superiority has been repeatedly verified in various tasks of computer vision, image processing. Deep learning network can be obtained by simply stacking many layers, where deep and shallow layers capture different features of an input image. Roughly speaking, shallow layer extracts low-level image features, such as color, texture, edge and contour of partial objects. While deep layers extract features of high-level semantic information (e.g., identifiable objects or parts of an object). This mechanism of deep learning is consistent with human vision perception system (HVS). From aspect of algorithm/theory principle, more and more works [1–4] proved that generalization ability increases along with more layers of a network.

In astronomy, with fast development of high resolution and high accuracy astronomical instruments, modern astronomy has stepped into a "big data" era. Therefore, AI techniques, are highly demanded in modern astronomy study. On one hand, in spite of massive data, astronomical observation data has a very low value intensity, urgently need AI for data filtering/mining. On the other hand, astronomical observation also provides time series of routine monitoring data (e.g., magnetic field, X-ray, radio and extreme ultra-violent (EUV) wavelength monitoring to the Sun). By means of AI, we may find implicit causality of an astronomical event from historical data, for forecasting occurrence or development of the astronomical event in the future, e.g., solar flare forecasting.

1.2   Big Data of Astronomy

From 1970s, human has launched many satellites for various missions of deep space exploration, solar, lunar and Mars explorations. The well-known Solar Dynamics Observatory (SDO) [5], Solar and Heliospheric Observatory (SOHO) [6] and Solar TErrestrial RElations Observatory (STEREO) [7] provides plentiful observation data of high accuracy, resolution and cadence to astronomers around the world. Meanwhile, large ground-based telescopes have been developing more and more in the world, such as SKA [8], Arecibo [9], ALMA [10], MUSER [11] and FAST [12]. The newly developed telescopes are getting better and better in both sensitivity and resolution; meanwhile, they create huge volume of observation data. In spite of massive data of astronomical observation, the density of data value is very low (Roughly estimating, less than 5% of solar radio observation is valuable to solar physicists). In this sense, only binary classification of observation data would be of great importance for data processing, reducing human labors dramatically. To the end, we take advantage of deep learning to classify massive data automatically.

The ground-based solar telescopes include magnetograph, Hα and radio telescopes. The former two are usually optical single-dish telescope. The last one has both single-dish and dish array (MUSER [11], Nobeyama [13] and SSRT [14] radioheliograph) for recording total solar irradiance and imaging the Sun, respectively. MUSER is a solar dedicated interferometric array consisting of 100 dishes. It images the Sun at Ultra-wideband (0.4–15 GHz) with 528+64 frequency channels, less than 200 ms time sampling rate, resulting in 3–5 TB raw data every day. The big data has caused a big challenge to data archiving, classification, processing and storage. Nowadays, it is impossible to manipulate big data manually, so automatic data archiving by help of AI is highly demanded.

1.3 Deep Learning

AI was raised by Dartmouth conference in 1956. It has undergone three waves in history. In the first wave, it developed logical reasoning successfully for theorem proving. The second wave started from 1976, lasted for 30 years. During that time, expert systems were largely exploited. The third wave was characterized by deep learning, which started from 2006.

Deep learning is essentially neural network with more than one hidden layers. In the past, for lack of data, computing power, only shallow neural network was used before 2006. Thanking to big data, high performance computing (especially, GPU computing) and training skills (e.g., back propagation (BP)), deep learning got its outbreak since 2006. It has been verified that deep network is superior to shallow network for exploring high-level semantic information which is more close to human cognition. For this reason, deep learning has brought a big breakthrough in image classification, object detection, machine translation and NLP.

Nowadays, with the availability of massive data, deep learning [15] has been extensively explored to solve many traditional tasks of recognition, classification, regression and clustering. The methods of deep learning, such as convolutional neural networks (CNN) [16, 17], auto-encoder (AE) [18], deep belief networks (DBN) [19–21], have demonstrated state-of-the-art performance in a wide variety of tasks, including visual recognition [22], audio recognition [23], and NLP [24]. These methods can directly learn representative features from input data without need of prior knowledge for extracting hand-crafted features. This advantage of deep learning is especially beneficial to the tasks that we were in the absence of solid domain knowledges, or concerned physical principles are so complicated that we cannot reason or model them precisely.

Convolutional neural network (CNN) is the most successful in the family of deep learning. In a CNN, convolution operation is confined in a local patch, simulating local receptive field of human visual system (HVS). In addition, the parameters of convolution kernel are shared across an entire image, so only small number of parameters are learnt regardless of image resolution. The success of CNN brought a family of deep learning models, including AlexNet [25], GoogleNet [26], VGGNet [17], ResNet [27] and DenseNet [28], achieving big success of image classification and recognition. ImageNet Large Scale Visual Recognition Competition (ILSVRC) [29] was a very famous international competition of image classification and recognition. Before 2012, it got the best result of 26% classification error by using hand-crafted features and shallow models. From 2012, all of the competitors adopted deep learning to build their models, gradually reducing error rate from 26% to 2%. Auto-encoder is a generative model, which principle is straightforward, i.e., compressing input signal to get its compressed representation, and then decompressing it to reconstruct signal. During this process, auto-encoder can acquire essential feature representation of input signal and compress redundant information. Recently, generative adversarial network (GAN) [30] becomes very popular. It is established on zero-sum game theory. Through adversarial learning between a generator and a discriminator alternately, one can finally acquire a good generator for image generation. The models above mentioned are all static, implying that only current input contributes to current output, without exploiting the context of sequential inputs. Recurrent neural network (RNN) is a dynamic model for processing sequential inputs by providing a loop unit, so that history status of network can be kept for current output. In the family of RNN, long short-term memory (LSTM) [31] network gets people's favour due to its efficiency and easy-to-use. LSTM controls information flow via three gates, achieving good performance on time series modeling.

1.4   Deep Learning in Solar Astronomy

AI is mainly for the purpose of classification and generation. Date archiving/selection figures out interesting events from daily observations, which is in essence a classification task, e.g., finding new starts in sky survey, and detecting solar bursts in solar observation. In solar astronomy, we employ classification models of deep learning to classify solar radio spectrums, detect active region, sunspot, filament, solar flare, coronal hole and etc. Image generation is closely concerned with visualization of observed data, image reconstruction and image enhancement. The classification and generation mentioned above are all concerned with static image processing. In astronomy, another important application of AI is to predict future from history data, namely forecasting, such as solar flare forecasting. Forecasting is also a classification task in essence. Specifically, it makes a decision for the future with the current or past inputs. In most cases, it utilizes time series data to establish dynamic model.

To modern astronomy, big data is a great challenge regarding storage, bandwidth and computing capability. Meanwhile, mining knowledge from massive data raises high demand for machine learning. The value intensity of astronomical observation data is very low, so machine learning is employed to mine useful data. Then, the mined information goes through further analysis and study. Classification, together with recognition, detection, clustering and forecasting are collectively known as data mining, mining valuable knowledge from massive data. Shallow machine learning, such as Bayesian learning, maximum likelihood estimation (MLE), maximum a posterior estimation (MAP) and support vector machine (SVM), were used in the past. From 2006, deep learning has swept all fields of computer vision, NLP, machine translation, image processing, etc., so it has been extensively explored for all kinds of tasks of data mining.

In this review, some applications of deep learning in solar astronomy are presented and discussed. These applications can be categorized into three classes, classification, generation and prediction/forecasting, concerning massive solar data selection/filtering, solar data imaging and image enhancement, and solar burst forecasting. It has been witnessed that deep learning could significantly boosts scientific research of solar astronomy, big solar data mining and data-driven modeling of solar bursts.

References

1. Zhang J, Liu T, Tao D. 2018 An Information-Theoretic View for Deep Learning.
2. Novak R, Bahri Y, Abolafia DA, Pennington J, Sohl-Dickstein J. 2018 Sensitivity and Generalization in Neural Networks: an Empirical Study.
3. Canziani A, Paszke A, Culurciello E. 2017 An Analysis of Deep Neural Network Models for Practical Applications.
4. Neyshabur B, Tomioka R, Srebro N. 2014 In Search of the Real Inductive Bias: On the Role of Implicit Regularization in Deep Learning.
5. SDO Obseratory. [EB/OL]. https://sdo.gsfc.nasa.gov/ Accessed Oct. 31, 2020.
6. SOHO Obseratory. [EB/OL]. https://sohowww.nascom.nasa.gov/ Accessed Oct. 31, 2020.
7. Stereo Obseratory. [EB/OL]. https://stereo.gsfc.nasa.gov/ Accessed Oct. 31, 2020.
8. SKA Obseratory. [EB/OL]. https://www.skatelescope.org/the-ska-project/ Accessed Oct. 31, 2020.
9. ARECIBO Obseratory. [EB/OL]. https://www.almaobservatory.org/en/home/ Accessed Oct. 31, 2020.
10. ALMA Obseratory. [EB/OL]. https://www.almaobservatory.org/en/home/ Accessed Oct. 31, 2020.
11. Mingantu Solar Radioheliograph. [EB/OL]. http://klsa.bao.ac.cn/ Accessed Oct. 31, 2020.
12. FAST Telescope. [EB/OL]. https://fast.bao.ac.cn/ Accessed Oct. 31, 2020.
13. Nobeyama Radio Telescope. [EB/OL]. https://www.nro.nao.ac.jp/ Accessed Oct. 31, 2020.
14. Siberian Solar Radio Telescope. [EB/OL]. http://en.iszf.irk.ru/ Accessed Oct. 31, 2020.
15. Bengio Y et al.. 2009 Learning deep architectures for AI. *Foundations and trends® in Machine Learning* **2**, 1–127.
16. LeCun Y, Boser B, Denker JS, Henderson D, Howard RE, Hubbard W, Jackel LD. 1989 Backpropagation applied to handwritten zip code recognition. *Neural computation* **1**, 541–551.

17. Simonyan K, Zisserman A. 2015 Very Deep Convolutional Networks for Large-Scale Image Recognition. In *International Conference on Learning Representations*.
18. Vincent P, Larochelle H, Bengio Y, Manzagol PA. 2008 Extracting and composing robust features with denoising autoencoders. In *Proceedings of the 25th international conference on Machine learning* pp. 1096–1103. ACM.
19. Hinton GE, Salakhutdinov RR. 2006 Reducing the Dimensionality of Data with Neural Networks. *Science* **313**, 504–507.
20. Hinton GE, Osindero S, Teh YW. 2006 A fast learning algorithm for deep belief nets. *Neural computation* **18**, 1527–1554.
21. Hinton GE. 2012 A practical guide to training restricted Boltzmann machines. In *Neural networks: Tricks of the trade* pp. 599–619. Springer.
22. Sohn K, Jung DY, Lee H, Hero AO. 2011 Efficient learning of sparse, distributed, convolutional feature representations for object recognition. In *Computer Vision (ICCV), 2011 IEEE International Conference on* pp. 2643–2650. IEEE.
23. Mohamed Ar, Dahl GE, Hinton G et al.. 2012 Acoustic modeling using deep belief networks. *IEEE Trans. Audio, Speech & Language Processing* **20**, 14–22.
24. Collobert R, Weston J, Bottou L, Karlen M, Kavukcuoglu K, Kuksa P. 2011 Natural language processing (almost) from scratch. *Journal of Machine Learning Research* **12**, 2493–2537.
25. Krizhevsky A, Sutskever I, Hinton G. 2012 ImageNet Classification with Deep Convolutional Neural Networks. *Neural Information Processing Systems* **25**.
26. Szegedy C, Wei Liu, Yangqing Jia, Sermanet P, Reed S, Anguelov D, Erhan D, Vanhoucke V, Rabinovich A. 2015 Going deeper with convolutions. In *2015 IEEE Conference on Computer Vision and Pattern Recognition (CVPR)* pp. 1–9.
27. He K, Zhang X, Ren S, Sun J. 2016 Deep residual learning for image recognition. In *Proceedings of the IEEE conference on computer vision and pattern recognition* pp. 770–778.
28. Huang G, Liu Z, Van Der Maaten L, Weinberger KQ. 2017 Densely Connected Convolutional Networks. In *2017 IEEE Conference on Computer Vision and Pattern Recognition (CVPR)* pp. 2261–2269.
29. Russakovsky O, Deng J, Su H, Krause J, Satheesh S, Ma S, Huang Z, Karpathy A, Khosla A, Bernstein M, Berg AC, Fei-Fei L. 2015 ImageNet Large Scale Visual Recognition Challenge. *International Journal of Computer Vision (IJCV)* **115**, 211–252.
30. Goodfellow IJ, Pouget-Abadie J, Mirza M, Xu B, Warde-Farley D, Ozair S, Courville AC, Bengio Y. 2014 Generative Adversarial Networks. *ArXiv* **abs/1406.2661**.
31. Hochreiter S, Schmidhuber J. 1997 Long short-term memory. *Neural computation* **9**, 1735–1780.

Chapter 2
Classical Deep Learning Models

Abstract Deep learning has achieved a big success in computer vision, NLP, audio processing and machine translation. Accordingly, there have been a bunch of classical deep learning models designed for these tasks. In this chapter, convolutional neural network (CNN), LSTM, autoencoder (AE) and GAN are discussed briefly. These models are most efficient for processing image, time series (e.g., video, NLP) and image generation respectively, as the foundation of our proposed models in this book. Recently, more advanced deep learning models/principles have emerged, such as attention (e.g., non-local, squeeze and excitation (SE), global context (GC), and most popular transformer), graph convolution network (GCN), self-supervised learning and contrastive learning. They can further boost model performance, extend application filed and break the limits of lack of labelled data, noise data and etc.

Keywords Deep learning · Convolutional neural network (CNN) · Generative adversarial network (GAN) · Long short-term memory (LSTM) · Autoencoder (AE)

2.1 Convolutional Neural Network (CNN)

Fully connected neural network (FCN) receives an input, and transform it through a series of hidden layers, where each neuron is fully connected to all neurons in the previous layer. It does not fit for image since the huge number of parameters would lead to over-fitting. In fact, the study about mammalian visual system found that a visual neuron was only sensitive to a local region of an image instead of the whole image. This finding revealed that mammalian visual system had a local receptive field, inspired us to develop CNNs which is the most successful model for image processing. In a CNN, each neuron of a higher layer is only related to a small region of neurons instead of all neurons in the previous layer. A CNN diagram is shown in Fig. 2.1. With local receptive fields, we can extract elementary visual features such as oriented edges, end-points, corners (or similar features in speech spectrograms). These features are then combined together in the higher layers.

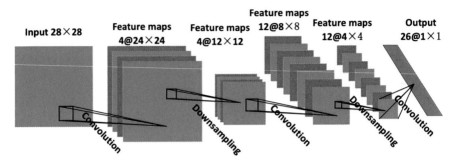

Fig. 2.1 Framework of convolution neural network (CNN)

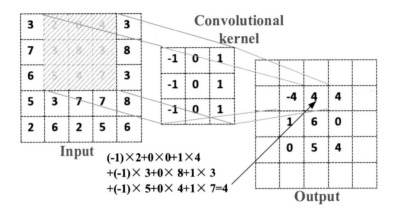

Fig. 2.2 Illustration of how convolution operation is performed

The ideas of local receptive field and weights sharing are achieved by introducing a convolution kernel, i.e., a small filter (e.g., 3×3, 5×5) sliding over input image pixel by pixel as demonstrated in Fig. 2.2. By this way, the number of parameters is determined by the size of convolution kernel instead of image resolution. In the early days, it was difficult to train a deep CNN without over-fitting due to lack of training data and computational ability. With the boom of big data and GPU computing in the last decade, CNNs became very successful. In addition, ILSVRC boosted the fast development of CNNs on image classification/generation, object detection/recognition. We can noticed that CNNs swept all fields of computer vision after the first success of CNN, namely AlexNet [1], in ILSVRC 2012.

A CNN is composed of convolution, down-sampling and fully connected layers. The convolution operation mimics local receptive field in biology, characterized by local receptive field and weight sharing, which can greatly compress network parameters. Down-sampling, which is also named pooling, is the sub-sampling of images. It is used to reduce the amount of data while still retaining useful information. After stacking multiple convolutions and pooling layers, one or more fully connected layers are connected to realize classification/regression.

2.2 Long Short-Term Memory (LSTM) Network

In the family of deep learning, most of them, such as FCN, CNN, GAN and auto-encoder, give output only from the current input. They are all static models without utilizing temporal context information of sequential inputs. However, for the tasks of video processing, language modeling, machine translation, speech recognition, reading comprehension, the context of input is of great importance for recognition and understanding. From aspect of human understanding, we don't start our thinking from scratch every second as we read a paper. Instead, we understand each word based on our understanding of previous words. We don't throw everything away and start thinking from scratch again. However, static models process inputs separately, output at every input. They do not tell how the history of input informs current status and output. RNN was raised to resolve this problem.

In Fig. 2.3, a diagram of RNN is illustrated, where a chunk of neural network, A, receives an input x_t and outputs h_t. There is a loop in an RNN, keeping history input for current output. A loop allows information to be passed from one step of the network to the next. An RNN can be thought of as multiple copies of the same network, each passing a message to its successor. It can be unrolled into a chain-like normal neural network as shown in Fig. 2.4 if we unroll the loop. This chain-like nature reveals that it is intimately related to sequences and lists. It is a natural architecture of neural network for sequential inputs, exploring correlations and interactions among adjacent inputs. In the last few years, a variety of applications of RNN, including speech recognition [2, 3], language modeling [4, 5], machine translation [6] and image captioning [7, 8], have achieved incredible success.

RNN keeps history inputs for better informing the understanding of current status, e.g., reading comprehension. In practice, RNN performs well for exploring short-term dependency, while inferior for long time dependency. Thus, LSTM, as an excellent representative of RNN, was raised for well processing both long-term and short-term dependencies. Both RNN and LSTM repeat a basic unit to form a chain-structure network. Compared to RNN, the basic unit of LSTM, namely LSTM cell, has more elaborate structure as shown in Fig. 2.5, consisting of four nonlinear layers. The cell status is the key of LSTM, kept in the horizontal axis of Fig. 2.5.

Fig. 2.3 Recurrent neural network with a loop

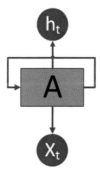

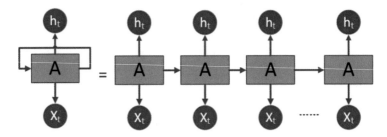

Fig. 2.4 Unrolled recurrent neural network

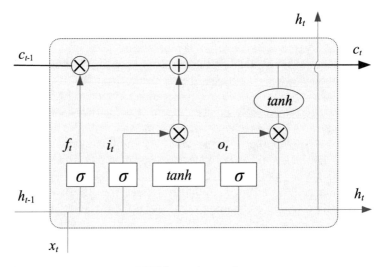

Fig. 2.5 Long short-term memory (LSTM) neural network

LSTM controls information flow through three gates, namely input gate, forget gate and output gate, enabling signal to pass the cell selectively, achieving information persistence and inhibiting. Intuitively, forget gate controls the extent to which a value remains in the cell, input gate controls the extent to which a new value flows into the cell and output gate controls the extent to which the value in the cell is used to compute the output activation of the cell.

In a LSTM cell, the mapping function from an input sequence $x = (x_1, x_2, \cdots, x_T)$ to an output sequence $h = (h_1, h_2, \cdots, h_T)$ is precisely specified by

$$f_t = \sigma(W_{fx}x_t + W_{fh}h_{t-1} + b_f) \tag{2.1}$$

$$i_t = \sigma(W_{ix}x_t + W_{ih}h_{t-1} + b_i) \tag{2.2}$$

$$o_t = \sigma(W_{ox}x_t + W_{oh}h_{t-1} + b_o) \tag{2.3}$$

$$c_t = f_t * c_{t-1} + i_t * \tilde{c}_t$$
$$\tilde{c}_t = \tanh(W_{cx}x_t + W_{ch}h_{t-1} + b_c) \tag{2.4}$$

$$h_t = o_t * \tanh(c_t) \tag{2.5}$$

where σ is logistic sigmoid function, and i, f, o and c are the input gate, forget gate, output gate, cell and cell input activation vectors, respectively, all of them are the same size as the cell output activation vector h. The W represent weight matrixes and the b represents bias vectors. The operator "$*$" represents the element-wise vector product.

The first step in LSTM is to decide what information we are going to throw away from the cell state. This decision is made by a sigmoid layer called the "forget gate". As described in (2.1), it looks at h_{t-1} and x_t, and outputs a number between 0 and 1 for each element in the cell state C_{t-1}. "1" represents that we completely keep the history of that element while "0" represents that we completely get rid of the history of that element. The second step of LSTM is to decide what information we will update into the cell state. This decision is made by a sigmoid layer called the "input gate", it given in (2.2). It looks at h_{t-1} and x_t, and outputs a number between 0 and 1. Next, it will be imposed on a *tanh* layer which creates a vector of new candidate values, $\tilde{c}_t$, that could be added to the new state c_t as given in (2.4). Therefore, input gate decides how much each element of state (candidate state) is updated.

After the two steps above, cell state c_t is updated by both old state c_{t-1} and a new candidate state ($\tilde{c}_t$) via (2.4), where old state c_{t-1} is scaled by forget gate f_t, indicating how much we forget the history, candidate state ($\tilde{c}_t$) is scaled by input gate i_t, controlling how much each state value of ($\tilde{c}_t$) is updated into the new state.

Finally, we decide what we are going to output. This output will be based on current cell state, but will be a filtered version as described in (2.5). First, we run a sigmoid layer which decides what parts of the cell state we are going to output. This sigmoid layer is called "output gate", given by o_t in (2.3). Then, we put the cell state c_t through *tanh*, and multiply it by o_t, getting the final output h_t.

2.3 Generative Adversarial Network (GAN)

GAN [9, 10] is an excellent representative of generative deep learning (DL) models. It has been extensively investigated in image reconstruction, image denoising, image synthesis, super-resolution and etc. GAN is comprised of a generator and a discriminator. The generator makes fake image close to real one/ground-truth, while the discriminator distinguishes between fake image and real one as shown in Fig. 2.6. Repeating adversarial learning between them, a powerful generator can be learned, generating image very close to real one. The principle of GAN is originated

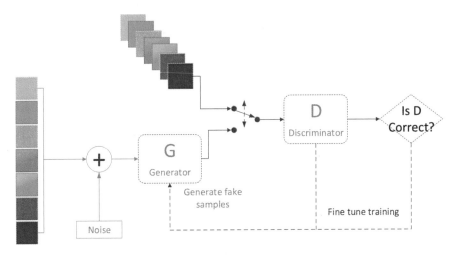

Fig. 2.6 Diagram of conditional GAN network

from the well-known zero-sum game, which is mathematically represented by

$$G^* = \arg\min_{G}\max_{D} \mathcal{L}_{GAN}(G, D)$$

$$\mathcal{L}_{GAN}(G, D) = \mathbb{E}_y[\log D(y)] + \mathbb{E}_z[\log(1 - D(G(z)))]$$

(2.6)

where D represents a detector, G represent a generator, y is a real image and $G(x, z)$ is a fake image. In (2.6), y is coming from a distribution of real data, and z is coming from a random noise (e.g., Gaussian noise). For optimizing D, we expect the larger $D(y)$ on the real data and the smaller $D(G(z))$ on the fake data generated by the generator G. While for optimizing G, we expect that it can generate enough realistic sample $G(z)$ to cheat D successfully. During such an adversarial process, D and G are optimized alternatively.

As (2.6) indicated, a general GAN only discriminate fake and true of the output. However, most of image processing tasks, e.g., well-known image-to-image translation [11], require the correspondences between inputs and outputs besides discriminating fake and true. For this purpose, conditional GAN (cGAN) was proposed, which is described by

$$G^* = \arg\min_{G}\max_{D} \mathcal{L}_{cGAN}(G, D)$$

$$\mathcal{L}_{cGAN}(G, D) = \mathbb{E}_{x,y}[\log D(x, y)] + \mathbb{E}_{x,z}[\log(1 - D(x, G(x, z)))]$$

(2.7)

where $D(x, y)$, $D(x, G(x, z))$ indicates that D needs not only distinguish the real and the fake, but also tell the correspondence between them. In [11], Phillip Isola et.al. described a cGAN model for image-to-image translation, namely pix2pix.

2.4 Autoencoder (AE)

An auto-encoder (AE) is a type of deep neural network for encoding input data to get its compressed representation, and then decoding this presentation to recover the input as much as possible. A basic AE is demonstrated in Fig. 2.7, where the encoder provides the compressed code, while the decoder produce the recovery of the input from this compressed code. In an AE, the encoding and decoding processes are jointly optimized to regenerate the input from the compressed code. The purpose of AE is typically dimensionality reduction, compression and denoising, by compressing insignificant "noises" and preserving "real" signal of input data.

AE is also a generative model, representing the input data by the functions in a neural network space, which makes the input data poss some special properties. It consists of an encoder and a decoder, where the encoder compresses the input data into a compressed code, while the decoder decompresses this code into a reconstruction of the input. Typically, AEs are forced to reconstruct the input approximately, preserving only the most relevant aspects of the input data.

As an unsupervised learning, AE can be stacked on any neural network for model pre-training in case of lack of labelled data. We can first train a stacked AE using unlabelled data, then use the lower layers of the AE to create a deep neural network for the actual task, refining it using the labelled data.

There have been many variants of AE, such as denoising AE, sparse AE, contrastive AE, adversarial AE and variational AE:

Denoising Autoencoder (DAE) It learns useful features by adding noise to the original input, then training itself to recover the original, noise-free inputs. This

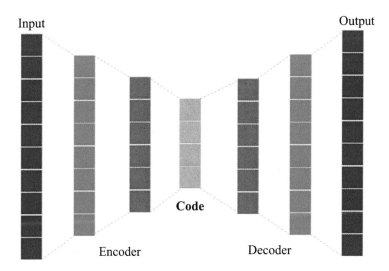

Fig. 2.7 The sketch of an auto-encoder

prevents the autoencoder from trivially copying its inputs to its outputs, so it could find patterns in the data. The noise can be pure Gaussian noise added to the inputs, or it can be randomly switched off inputs, just like in dropout.

The process for training a DAE is as follows:

(1) The initial input x is corrupted into $\tilde{x}$ through stochastic mapping $\tilde{x} \sim q_D(\tilde{x}|x)$.
(2) The corrupted input $\tilde{x}$ is then mapped to a hidden representation with the same process of the standard autoencoder, $h = f_\theta(\tilde{x}) = s(W\tilde{x} + b)$.
(3) From the hidden representation the model reconstructs $z = g_{\theta'}(h)$.

The model's parameters θ and θ' are trained to minimize the average reconstruction error over the training data, specifically, minimizing the difference between z and the original uncorrupted input x.

Sparse Autoencoder (SAE) Sparsity is a kind of constraint that could lead to good feature extraction in image processing. By adding a term to the cost function, SAE is forced to reduce the number of active neurons in the coding layer. The more sparse a layer, the better feature extraction. The sparsity can be realized in different ways by formulating the penalty terms. Kullback-Leibler (KL) divergence is one of common penalty term, given by

$$\sum_{j=1}^{s} KL(\rho||\hat{\rho}_j) = \sum_{j=1}^{s} \left[\rho \log \frac{\rho}{\hat{\rho}_j} + (1 - \rho) \log \frac{1 - \rho}{1 - \hat{\rho}_j} \right] \qquad (2.8)$$

where s and m represent the number of hidden nodes in a hidden layer and the number of training samples respectively, the sparsity parameter ρ is close to zero, $\hat{\rho}_j$ is defined by

$$\hat{\rho}_j = \frac{1}{m} \sum_{i=1}^{m} [h_j(x_i)], \qquad (2.9)$$

which represents the average activation of the hidden unit j over all training samples, where $h_j(x_i)$ identifies the input value that triggered the activation. To encourage most of the neurons to be inactive, $\hat{\rho}_j$ needs to be close to ρ. $KL(\rho||\hat{\rho}_j)$ is the KL-divergence between a Bernoulli random variable with mean ρ and a Bernoulli random variable with mean $\hat{\rho}_j$. Another way to achieve sparsity is by imposing L1 or L2 regularization terms on the loss function. For example, in case of L1 regularization term, the loss function is written as

$$\mathcal{L}(x, z) + \lambda \sum_{i} |h| \qquad (2.10)$$

Contrastive Autoencoder (CAE) It is encouraged to map a neighborhood of input points to a smaller neighborhood of output points, forcing the model to learn an encoding robust to the slight variations of the inputs by adding an explicit regularizer in loss function. This regularizer corresponds to the Frobenius norm of the Jacobian matrix of the encoder activations with respect to the input, encouraging the model to learn useful information about the training distribution. The final objective function is given by

$$\mathcal{L}(x, z) + \lambda \sum_i ||\nabla_x h_i||^2 \tag{2.11}$$

Variational Autoencoder (VAE) VAE is a probabilistic model, indicating that its output is partly determined by chance. More importantly, it is generative model, meaning that it can generate new instances that look like it was sampled from the training set. Instead of producing a coding for a given input directly, the encoder produces a mean coding μ and a standard deviation σ. The actual coding is then sampled randomly from a Gaussian distribution $G(\mu, \sigma)$. After that, the decoder just decodes the sampled coding normally. The sketch of a VAE is demonstrated in Fig. 2.8.

A VAE tends to produce codings that look as if they were sampled from a simple Gaussian distribution. The cost function of VAE is composed of two parts. The first is the usual reconstruction loss that pushes the autoencoder to reproduce its inputs. The second is the latent loss that could push the codings to gradually migrate within the coding space (latent space) to occupy a roughly (hyper)spherical region that looks like a cloud of Gaussian points. The consequence is that we can easily sample a new instance from this Gaussian distribution. For this purpose, the KL divergence between the target distribution (Gaussian distribution) and the actual distribution of the codings is employed as the latent loss.

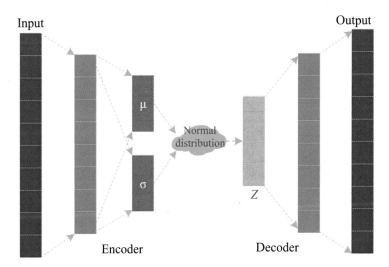

Fig. 2.8 The architecture of a Variational autoencoder (VAE)

Given an input dataset $\mathbf{x}$ with an unknown probability function $P(\mathbf{x})$ and a multivariate latent encoding vector $\mathbf{z}$, the objective is to transfer the data into a distribution $p_\theta(\mathbf{x})$,

$$p_\theta(\mathbf{x}) = \int_{\mathbf{z}} p_\theta(\mathbf{x}, \mathbf{z}) d\mathbf{z} = \int_{\mathbf{z}} p_\theta(\mathbf{x}|\mathbf{z}) p_\theta(\mathbf{z}) d\mathbf{z}, \qquad (2.12)$$

where, $p_\theta(\mathbf{x}|\mathbf{z})$ describes a network for generating $\mathbf{x}$ from $\mathbf{z}$, θ is the network parameter. With an assumption that $\mathbf{p}(\mathbf{z})$ is a normal distribution, generating $\mathbf{x}$ can be interpreted as randomly sampling an instance $\mathbf{z}$ from a normal distribution to calculate $\mathbf{x}$.

Adversarial Autoencoder (AAE) AAE blends the autoencoder with the adversarial loss given by GAN, as demonstrated in Fig. 2.9. It uses a similar concept with VAE except that it uses adversarial loss to regularize the latent code instead of the KL-divergence of VAE. In VAE, KL-divergence is used to match the encoded latent code into a normal distribution, which was replaced by adversarial loss where an additional discriminator component is augmented, and the encoder acts as the generator. Unlike basic GAN where the generator generates fake images, the generator of AAE generates latent codes to cheat the discriminator into believing that the latent codes are sampled from the normal distribution. The discriminator of AAE predicts whether a given latent code is generated by the autoencoder (fake) or a random vector sampled from the normal distribution (real).

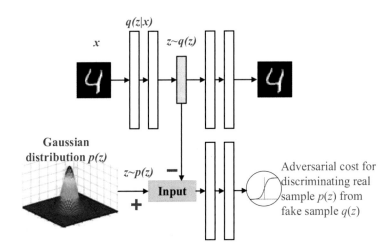

Fig. 2.9 The architecture of an Adversarial autoencoder (AAE)

References

1. Krizhevsky A, Sutskever I, Hinton G. 2012 ImageNet Classification with Deep Convolutional Neural Networks. *Neural Information Processing Systems* **25**.
2. Graves A, Mohamed Ar, Hinton G. 2013 Speech recognition with deep recurrent neural networks. In *Acoustics, speech and signal processing (icassp), 2013 IEEE international conference on* pp. 6645–6649. IEEE.
3. Graves A, Jaitly N. 2014 Towards end-to-end speech recognition with recurrent neural networks. In *International Conference on Machine Learning* pp. 1764–1772.
4. Sundermeyer M, Schlüter R, Ney H. 2012 LSTM neural networks for language modeling. In *Thirteenth annual conference of the international speech communication association*.
5. Mikolov T, Karafiát M, Burget L, Černocký J, Khudanpur S. 2010 Recurrent neural network based language model. In *Eleventh Annual Conference of the International Speech Communication Association*.
6. Cho K, Van Merriënboer B, Gulcehre C, Bahdanau D, Bougares F, Schwenk H, Bengio Y. 2014 Learning phrase representations using RNN encoder-decoder for statistical machine translation. *arXiv preprint arXiv:1406.1078*.
7. Chen L, Zhang H, Xiao J, Nie L, Shao J, Liu W, Chua TS. 2017 Sca-cnn: Spatial and channel-wise attention in convolutional networks for image captioning. In *2017 IEEE Conference on Computer Vision and Pattern Recognition (CVPR)* pp. 6298–6306. IEEE.
8. Ren Z, Wang X, Zhang N, Lv X, Li LJ. 2017 Deep reinforcement learning-based image captioning with embedding reward. *arXiv preprint arXiv:1704.03899*.
9. Goodfellow I, Pouget-Abadie J, Mirza M, Xu B, Warde-Farley D, Ozair S, Courville A, Bengio Y. 2014 Generative adversarial nets. In *Advances in neural information processing systems* pp. 2672–2680.
10. Xu L, Yan Y, Sun W. 2018 Solar Image Deconvolution by Generative Adversarial Network. In *AGU Fall Meeting Abstracts*.
11. Isola P, Zhu JY, Zhou T, Efros AA. 2016 Image-to-Image Translation with Conditional Adversarial Networks. *arXiv e-prints*.

Chapter 3
Deep Learning in Solar Image Classification Tasks

Abstract The exponential increasing of data being collected in astronomy has raised a big data challenge. Mining valuable information timely and efficiently from massive raw data is highly demanded. Even simple binary classification of collected raw data is of great importance, reducing the burden of the following data processing. Inspired by the success of image classification with deep learning, we investigated solar radio spectrum classification using deep learning, including the premier deep belief network (DBN), the most popular convolutional neural network (CNN) and long short-term memory (LSTM) network. For model training, a database of solar spectrum was established and published to the public. As far as we know, it is the first one in the world. The database contains 8816 spectrums with different image patterns which represent different solar radio mechanisms. Then, each spectrum was given a label by the invited experts of solar radio astronomy.

Keywords Solar radio spectrum · Image classification · Deep belief network (DBN) · Multimodality · Structured regularization

3.1 Solar Radio Spectrum

Solar radio spectrometer, as one of the most common instruments of solar observation, is widely used on the ground to record total solar radio radiation flux, providing an index of long-term solar activity level. In addition, a spectrometer has multiple frequency channels. Thus, it could generate a two-dimensional image within a certain time interval, namely spectrum. In case of solar burst, solar radio radiation would have a dramatic amplification, corresponding to a jump over individual frequency channels, or an abrupt brightness of a region in spectrums. Figure 3.1 demonstrates some typical spectrums, such as "pulse", "drifting", "zebra" and "fiber" which are named after their shapes, where each spectrum displays a distinctive image pattern, indicating a distinct physical mechanism.

For model training, we collected spectrums of Solar Broadband Radio Spectrometer (SBRS) of China [1] to establish a spectrum database. The SBRS, consisting of five "component spectrometers" and covering a wide frequency range from

© The Author(s), under exclusive license to Springer Nature Singapore Pte Ltd. 2022   19
L. Xu et al., *Deep Learning in Solar Astronomy*, SpringerBriefs in Computer Science, https://doi.org/10.1007/978-981-19-2746-1_3

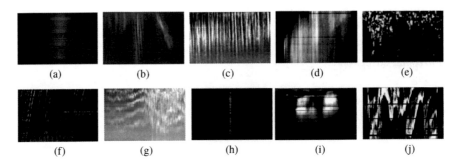

Fig. 3.1 Examples of solar radio spectrums. (**a**) Simple. (**b**) Type III. (**c**) Pulse. (**d**) Drifting pulse. (**e**) Spikes radiation. (**f**) Fiber. (**g**) Zebra. (**h**) Isolation pulse. (**i**) Patches. (**j**) Lace

0.7-7.6 GHz, monitors solar radio activity continuously, therefore accumulates massive data. Since solar burst event is very sparse among all data (less than 5% of all recorded data), only binary classification is of great importance for daily data archiving. Through binary classification, we can firstly figure out bursts from massive data, saving more than 95% human labors for the following data analysis. In this section, a bunch of deep learning models for solar radio spectrum classification are presented and discussed. In addition, we already have plentiful knowledges about the correspondences between fine structures of spectrums and their physical mechanisms. Thus, to collect various kinds of spectrums through classification model, we can do more on scientific research of solar radio astronomy.

3.2 Pre-Processing of Solar Radio Spectrum

The raw solar radio data captured by SBRS contains the flux values of radio radiation as well as the observation time. Although the captured data covers all the information of the solar radiation, it is hard for the viewers/researchers to judge or determine whether the solar burst happens or not and in what level the solar burst is. For easy understanding, raw data captured by SBRS is first converted into images for visualization.

SBRS contains several channels to monitor the solar burst in different frequencies. Therefore, the signal sensed from each channel will be treated individually. In total, there are 120 channels working toward the solar radio information captured at the same time. Moreover, each captured file contains both left and right circular polarization parts, which should be separated and processed individually for visualization and further processing. We extract the captured data from each channel as a row vector, which is stored according the sensing time. Afterwards, all the vectors from the 120 channels will be assembled together according the frequency values to form a solar radio spectrum, which is used for visualization and further processing. As there are 120 channels and 2520 sensing time points in

8ms recorded file, the final resolution of the converted image is 2520×120. One sample image is illustrated in Fig. 3.2a.

After the conversion process, it can be observed that there are numbers of horizontal-stripes-like interference signal almost in each picture, as illustrated in Fig. 3.2a. This phenomenon is named as the channel effect in solar radio observation, which is caused by different gains of different channels. It can be observed in Fig. 3.2a that each channel produces the signal of the same magnitude. Therefore, clear horizontal lines can be easily detected from the captured solar radio spectrum. The channel effect may hide the presentation of bursts. To eliminate the channel effect, we propose one method for channel normalization, which is formulated by

$$g = f - f_{LM} + f_{GM}, \tag{3.1}$$

where f is the constructed image, g is the image after performing the channel normalization, f_{LM} and f_{GM} denote the local mean and global mean values, respectively. The local mean f_{LM} is the mean of the pixels in a frequency channel. f_{GM} accounts for the mean of a whole image. f_{LM} is to alleviate the effect of uneven channel gain resulting in horizontal-stripes-like interference, while the f_{GM} compensates each pixel value by adding a global background. The solar radio spectrum after channel normalization is illustrated in Fig. 3.2b. It can be observed that channel normalization removes horizontal-stripes-like noise successfully and the solar radio burst can be easier detected.

After imaging and normalization, we obtain spectrums of 2520×120 as shown in Fig. 3.2. It can be observed that neighboring columns are highly similar, indicating high redundancy. In a spectrum, each row gives solar radio flux at a certain frequency channel over a certain time interval. It can be regarded as an observation of a discrete stochastic process. Thus, each column represents a random variable. In probability and statistics, correlation of a stochastic process can be measured by correlation coefficient [10] [11] γ, which is defined as

$$\gamma(n_1, n_2) = \frac{\Phi[(x(n_1 - \mu_1)(x(n_2 - \mu_2)]}{(\sigma_1 \dot{\sigma}_2)} \tag{3.2}$$

where $x(n_1)$ and $x(n_2)$ represent two random variables at n_1 and n_2 with ensemble average, μ_1 and μ_2, and standard deviations, σ_1 and σ_2, and Φ represents an ensemble average operator. Thus, $\gamma(n, n + \tau)$ is a function of τ n, namely, it is not a stationary stochastic process. Consequently, for a given τ, $\gamma(\tau)$ is a random variable instead of a constant. We calculate $\gamma(\tau)$ with different values of τ. The results are illustrated in Table 3.1.

From Table 3.1, the rows have high correlation, while the columns are less correlated. Therefore, original spectrum can be down-sampled to remove the redundancy. We down-sample original spectrums into 75×30 by using nearest neighbor sampling method. It can be observed that image characteristic varies little comparing to original spectrums, as illustrated in Fig. 3.3.

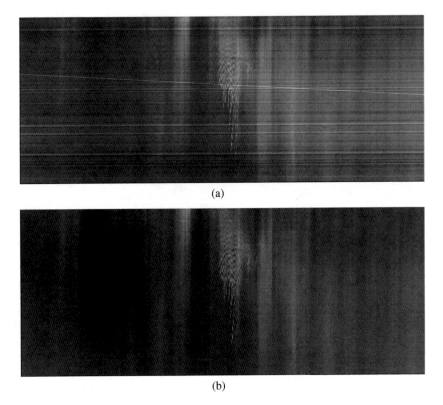

(a)

(b)

Fig. 3.2 Solar radio image before and after channel normalization. The horizontal axis denotes the sampling time, while the vertical axis indicates the frequency channel. (**a**) Solar radio image before channel normalization. (**b**) Solar radio image after channel normalization

Table 3.1 Correlation coefficients of solar radio images

$\gamma(\tau)$		$\gamma(\tau)$	
$\tau = 1$	0.9978	$\tau = 1$	0.8343
$\tau = 10$	0.9829	$\tau = 2$	0.7738
$\tau = 20$	0.9771	$\tau = 3$	0.7660
$\tau = 30$	0.9753	$\tau = 4$	0.7096
$\tau = 40$	0.9750	$\tau = 5$	0.6650

Due to the difference of calibration values, the spectrums have large gap of average gray, which may mislead neural network to learn gray difference instead of different image patterns of spectrums. Different image patterns of spectrums indicate different solar activity phenomena. Thus, image normalization is implemented over the whole dataset. It is different from the channel normalization given in (3.1). For image normalization, the local mean $f_L M$ is calculated over each spectrum, while the global mean $f_G M$ accounts for the average of local means over the whole dataset.

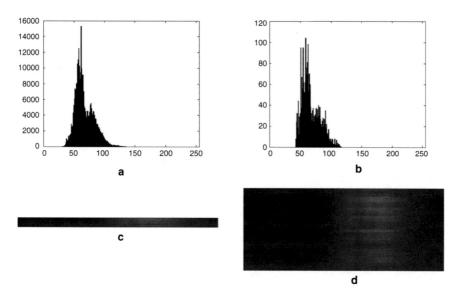

Fig. 3.3 The solar radio image and the histogram before and after down-sampling. (**a**) Histogram before down-sampling. (**b**) Histogram after down-sampling. (**c**) Original solar radio image. (**d**) Down-sampled solar radio image

After image normalization, image enhancement is further implemented before feeding into network. From our experience, most of the representative information of a spectrum is provided by the pixels around the mean of a spectrum, $\pm 30 + \overline{f}$. The rest contains more noise. To make spectrum more representative, a linear amplification is supplemented to compensate the pixels within $(\overline{f}\text{-}30, \overline{f}\text{+}30)$. Like general image, spectrum is also subject to a Gaussian distribution by examining histogram of spectrum. However, the mean of each spectrum varies greatly, leading to the situation that distribution of training set cannot be accurately modeled by a Gaussian distribution. In this situation, neural network would tend to capture overall grayscale of a spectrum, ignoring what a spectrum really represents, which mostly lies in the texture pattern of a spectrum. To prevent it from happening, a bilinear enhancement processing is further implemented after the preceding normalization. In Fig. 3.4, histograms of raw dataset, after normalization and after enhancement are illustrated. It can be observed that a more general Gaussian-like distribution is derived after normalization and enhancement.

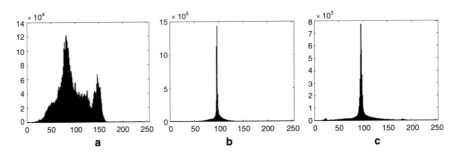

Fig. 3.4 Histograms of the solar radio image set. (**a**) Raw image set. (**b**) Normalized image set. (**c**) Enhanced image set

3.3 DBN for Solar Radio Spectrum Classification

Traditional shallow machine learning, e.g., support vector machine (SVM), needs hand-crafted image features for training classifier/regressor. Without enough priori knowledge, deep learning has great advantage over shallow models. It directly learns image features from an end-to-end network instead of hand-crafted features. DL has demonstrated state-of-the-art performance in a wide variety of tasks, including visual recognition, audio recognition, and NLP. The first DL model [2] that we have developed for spectrum classification is based on deep belief network (DBN). DBN [3] is a multi-layer, stochastic generative model stacking multiple restricted Boltzmann machines (RBMs) as shown in Fig. 3.5. RBM consists of one visible layer and one hidden layer, where the nodes are fully connected between two layers, without connections within each individual layer. Through RBM, neural network can achieve self-learning without need of labels, so it is usually used for model initialization. For classification, a classification layer is finally stacked on the top of RBM layers. Providing labelled data, the pre-trained model is further refined by a supervised learning.

For optimizing a RBM, an energy function is defined by

$$E(v, h) = -\sum_{i=1}^{V}\sum_{j=1}^{H} v_i h_j \omega_{ij} - \sum_{i=1}^{V} v_i b_i^v - \sum_{j=1}^{H} h_i b_i^h \tag{3.3}$$

where v is the binary state vector of the visible nodes, h is the binary state vector of the hidden nodes, v_i is the state of visible node i, h_j is the state of the hidden node j, ω_{ij} is the real-valued weight between the visible node i, the hidden node j. b_i^v is the real-valued bias to visible node i, and b_j^h is the real-valuded bias to hidden node j. The joint distribution of the visible and hidden nodes is defined by

$$p(v, h) = \frac{e^{-E(v,h)}}{\sum_u \sum_g e^{-E(u,g)}} \tag{3.4}$$

Fig. 3.5 Deep belief neural network

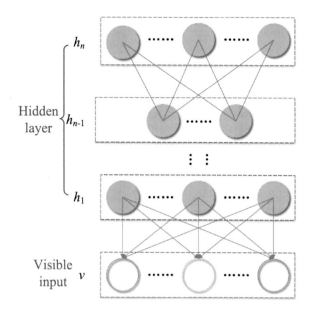

It can be observed that low energy would result in high probability, and vice versa. And also, the probability of an active visible node is independent of the states of the other visible nodes. Likewise, the hidden nodes are independent of each other. The property of the RBM makes the sampling extremely efficient, as one can sample all the hidden nodes simultaneously and then all the visible nodes simultaneously. However, the RBM given by (3.4) can only take binary-valued visible nodes, inhibiting its application on real-valued data.

For processing real-valued inputs, the Gaussian-Bernoulli RBM (G-RBM) was defined. Its energy function was modeled as

$$E(v, h) = -\sum_{i=1}^{V} \sum_{j=1}^{H} \frac{v_i}{\sigma_i} h_j \omega_{ij} - \sum_{i=1}^{V} \frac{(v_i - b_i^v)^2}{2\sigma_i^2} - \sum_{j=1}^{H} h_i b_i^h \qquad (3.5)$$

where v_i takes the real-valued activity of the visible node v_i. Each visible node adds a quadratic offset to the energy function where σ_i controls the corresponding width. Examining the binary visible and binary hidden node defined in (3.4), the G-RBM takes real-valued nodes as the input and output the binary nodes.

DBN is composed by more than one RBM layers. The weights of a higher layer are trained by fixing all the weights of the lower layers. Thus, it is trained greedily and sequentially. If the size of the second hidden layer is the same as the size of the first hidden layer, and the weights of the second is borrowed from the weights of the first, it can be proven that training the second hidden layer while keeping the first hidden layer's weights constant improves the log likelihood of the data under the model [4]. Figure 3.5 illustrates the multilayer DBN. The probability of the DBN assigns to a visible vector is defined as

$$p(v) = \sum_{h_1,\ldots,h_n} p(h_{n-1}, h_n) \prod_{k=2}^{n-1} p(h_{k-1}|h_k)p(v|h_1) \tag{3.6}$$

where n gives the number of hidden layers. DBN is demonstrated to be very helpful for hand-written digital numbers recognition, as well as unsupervised training [5]. In our work, it is employed to learn the representation of solar radio spectrum.

Due to the limited training samples, only one hidden layer is employed in our model. Then, a "I-H-C" structured neural network for spectrum classification is raised, where "C", "I" and "H" stand for classification, input layer and hidden layer respectively. The bottom layer of the network is a RBM and the top layer is a softmax layer. The objective function is given by $\hat{o} = \arg\max_o p(o|x; \Theta)$, where Θ includes all the parameters in the RBM and the softmax layer. For model training, standard contrastive divergence learning procedure is employed for pre-training.

The depth of network usually depends on the size of training set. For training a deep learning model, over-fitting is a big challenge in case of lacking training data. Fortunately, RBM can realize unsupervised learning, so it was well developed in network initialization and pre-training previously. Pre-training could alleviate the complexity of training task dramatically without need of massive labels, refraining from the risk of trapping in poor local optima. For training a RBM layer, standard contrastive divergence learning procedure is performed firstly. After pre-training, fine-tuning process is further implemented to make the network more specific for spectrum classification, where objective function takes the shape of a log-likelihood function. Furthermore, in order to prevent over-fitting, drop-out is employed. Typically, the outputs of neurons are set to zero with a probability of p in the training stage and multiplied with $1 - p$ in the test stage. By randomly masking out the neurons, drop-out is an efficient approximation of training many different networks with shared weights.

For spectrum classification, a classification layer with three output nodes is stacked on the top of the RBM layer, which takes output of RBM as input and outputs probabilities of the input falling into one of these three classes.

The experimental results given in Table 3.6 demonstrates that DBN can learn better representation of solar radio spectrum, and thus achieve higher accuracy of classification beyond the traditional SVM [6] coupled with principal component analysis (PCA) [7, 8].

3.4 Autoencoder for Spectrum Classification

Besides RBM, autoencoder (AE) is another discipline to realize unsupervised learning. It is optimized by setting output equal to input. In other words, it learns an approximation to identity function, so as to network output is identical to input.

The identity function seems a particularly trivial function. But, by placing certain constraints on network, such as the limited number of hidden units, interesting structure about the data can be learned. In our previous work [2], it has been applied to learn representation of spectrum successfully. In the literature, the variances of AE, such as denoising AE [9], stacked AE (SAE) [10] were also developed. However, these AEs treat the input equally, which cannot distinguish the characteristics between different input modalities well. Thus the interaction between different modality inputs cannot be captured.

In [5], an automatic dimensionality reduction method was proposed to facilitate the classification, visualization, communication, and storage of high-dimensional data through an adaptive, consisting of a multilayer "encoder" network to transform high-dimensional data into low-dimensional code and a "decoder" network to recover high-dimensional data from low-dimensional code. Using random weights as the initialization in these two networks, they can be trained together by minimizing the discrepancy between the original data and its reconstruction. Then, compressed representation of input data can be learned through an unsupervised learning. Inspired by these achievements, we can learn representation of spectrum better than [2], which would benefit the following classification, recognition, storage and analysis.

In [11] and [12], auto-encoder (AE) [9, 10, 13] was explored for spectrum classification. In addition, multimodal concept [14, 15] was employed to exploit the correlations among adjacent frequency channels, where each channel was regarded as one modality (Fig. 3.6). Specifically, the network was built on an AE with structured regularization (SR), to learn representation of spectrum. Then, it further classifies spectrums using a softmax layer.

The multimodal architecture takes different numbers and types of modalities as the input. The output will be the classification results, which not only considers the property of each modality but also accounts for the interactions between different modalities. The proposed multimodal learning architecture is built by stacking a softmax layer on the top of AE layer with structured regularization. Observing Fig. 3.7, the difference between single modality and multimodality lie in that each modality is trained separately in the low layers. It has proved that multimodal

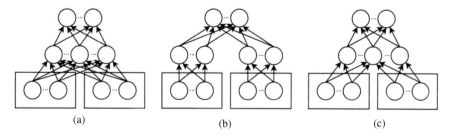

(a)   (b)   (c)

Fig. 3.6 Three models for multimodal input data learning. (**a**) Fully dense model. (**b**) Modality-sepcific model. (**c**) Group sparse model

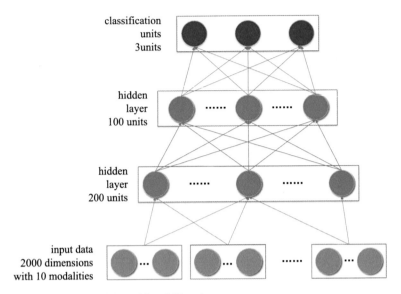

Fig. 3.7 The framework of the multimodal learning

learning method was superior to single modality of deep learning [2] from our evaluations over the established database described in Sects. 3.1 and 3.2.

In the proposed model, AE is used in each modality learning from input layer to the last hidden layer. All modalities are interacted through a SR among them. The global loss function of the proposed multimodal learning is written as

$$\hat{L} = \phi(f_{SR}(x_1, x_2, \ldots, x_m)) \tag{3.7}$$

where $\hat{L}$ is the network output, ϕ represents a non-linear function which is a softmax function in our model, and $f_{SR}()$ is a linear transform from the visual layer to the first hidden layer with SR. Using SR in AE, we get joint representations of the input signals through exploring their correlations.

3.4.1 Structured Regularization (SR)

SR is employed in AE with multimodal inputs inspired by Fawcett [16] and Jalali [17]. Each modality will be used as a regularization group separately for each hidden unit, similar to group regularization. This is different from conventional SR, where each input unit is treated equally, ignoring correlation between different modalities. Suppose $S_{r,i}$ as a $K \times N$ modality binary matrix, where K denotes the number of modalities and N indicates the number of units in the corresponding modality, SR is defined as

$$SR(W^{[1]}) = \sum_{j=1}^{M} \sum_{k=1}^{N} f_B(\max_i(S_{k,i}|W_{i,j}^{[1]}|) > 0) \qquad (3.8)$$

where f_B indicates a Boolean function that takes a value of 1 if its variable is true, and 0 otherwise. The regularization function in (3.8) performs a direct penalty on the number of modalities used for each weight.

3.4.2 Integrating SR with AE

By integrating SR with AE, the objective function for training the network is written as:

$$W^{[1]} = \arg\min_{W^{[1]}} \sum_{i=1}^{n^{([1])}} \|z_i^{[1]} - x_i^{[1]}\|_2^2 + \alpha \cdot SR(W^{[1]})$$

$$z_i^{[1]} = \sum_{j}^{k^{[1]}} \mu_j^{[1]} W_{i,j}^{[1]} \qquad (3.9)$$

where $z_i^{[1]}$ is the signal reconstructed by the decoder of AE. $n^{[1]}$ is the number of the input nodes including all the modality features, and $k^{[1]}$ is the number of the hidden nodes of the multimodal AE. $W_{i,j}^{[1]}$ is the weights of the multimodal AE by introducing SR. α is the parameter to balance the error and the regularization terms. By integrating SR into AE, the obtained representation y_i only connects to partial nodes in the first hidden layer. As shown 3.8, to minimize $SR(W^{[1]})$, $W_{i,j}^{[1]}$ should approach 0 as far as possible, leading to some nodes in the first hidden layer connected to only part of the nodes in the visual layer. AE with SR demonstrates that the multimodal network could distinguish different modalities and learn the correlations between them automatically.

3.4.3 Network Architecture

We propose an "I-H-C" structure network as shown in Fig. 3.7 for solar radio spectrum classification. "I" represents multimodal inputs, which is of the dimension of 2000 in our case. It should be pointed that these input nodes are arranged into 10 modalities, each of which is the size of 200. "H" de-notes the hidden layer, which consists of 200 nodes for the first hidden layer and 100 nodes for the second layer. "C" is defined as the classification nodes which calculates the probabilities of each input for the given classes. There are 3 outputs: "burst", "non-burst" and "calibration" in our application.

Input layer contains 10 modalities. Generally, modality indicates different type of data, e.g., audio, image, language, text and so on. In this application, frequency channels of a spectrum are regarded as modalities. There are three possible models for multimodal learning. A naive way of applying feature learning to multimodal data is to simply take the whole data vector as input to the model. This approach called "Fully dense model" may fail to learn associations between modalities with very different underlying statistics. In addition, it learns features prematurely, which can lead to overfitting. Instead of the fully dense model, modality-specific sparse model trains a first layer representation for each modality separately. This approach assumes that the ideal low-level features for each modality are purely unimodal, while higher-layer features are purely multimodal. So, this approach may work better for some problems where the modalities have very different basic representations, such as image and text. In our application, frequency channels of spectrums are treated as the modalities. These modalities have strong correlations between each other, which means that the learning of low-level correlations may lead to better features. Therefore, the group sparse model shown in Fig. 3.6c is employed, and a multimodal group regularization algorithm is proposed for learning the joint features of multiple channels. The AE with SR is employed to establish the multimodal learning network. Due to limited number of labeled solar radio spectrums, the number of layers and nodes cannot be too large. In this proposed network, only two hidden layers are employed for avoiding overfitting. A softmax layer is added on the top of the multimodal learning architecture realized by AE with regularization mentioned above. It takes the joint representations of multiple modalities as inputs, and outputs the classification results for each spectrum. The classification layer will determine the probabilities of an input spectrum regarding to given classes.

3.5 CNN for Solar Radio Spectrum Classification

Since solar radio spectrum is a special type of image, it can be expected that better image representation and classifier of spectrum can be learnt by using CNN. The first attempt to utilize CNN for spectrum classification was presented in [18]. As shown in Fig. 3.8, the proposed model consists of four convolution layers, four pooling layers, one fully connected layer and a softmax layer, which are stacked together to realize both feature extraction and classification. Experimental results in Table 3.6 indicate that the proposed CNN can achieve better performance beyond our previous effort of DBN [2] and autoencoder (AE) [12]. CNN is composed of convolution, down-sampling and fully connected layers. The theoretical basis of the convolution layer is mainly the concept of receptive field in biology, and local

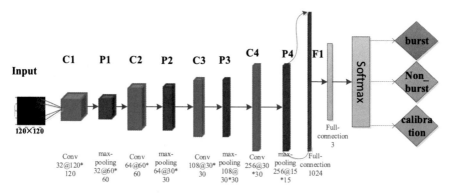

Fig. 3.8 Architecture of the proposed CNN for spectrum classification

receptive field and weight sharing are the common points between convolution and receptive filed, which can greatly reduce the parameters that neural network needs to train. Down-sampling, which is also named pooling, is the sub-sampling of images in fact. It is used to reduce the amount of data while still retaining useful information. By stacking the convolutions and the pooling layers, one or more fully connected layers can be formed, enabling higher-order inference capabilities.

Our proposed CNN model shown in Fig. 3.2 consists of four pairs convolution layers and corresponding pooling layers (C1-P1, C2-P2, C3-P3 and C4-P4) followed by a fully connected layer (F1). We use spectrum of the size 120×120 after preprocessing as the input of the network. C1 contains 1×5 patch filter, the purpose of which is to extracts the local features of the input data and constructs the feature maps in layer C1. Assuming 32 convolution kernels, we obtain 32 feature maps with the size 120×120 after C1. Then, these obtained feature maps are pooled in P1. Here, we use 2×2 pooling kernel. After pooling, the 120×120 feature maps are reduced to 60×60 feature maps.

C2-P2, C3-P3 and C4-P4 have the same structure as C1-P1 in the proposed model. We use the same activation function, Relu in all convolution layers. The number of feature maps of C2, C3, C4 are 64,128 and 256 respectively. The kernel size is 1×5 for C1-C3. Different from C1, C2 and C3, the kernel size is 1×3 for C4. After C4-P4, we can get 256 feature maps with the size of 8×8. Then, a fully-connected layer F1 with 1024 nodes is applied to output of C4-P4. Rectified unit is used as activation function, and dropout is with a probability of 0.75 in F1 to accelerate convergence and avoid excessive dependency on certain nodes. Finally, a softmax layer is stacked on the top of the network for the purpose of classification. For clearly understanding of the data flow of the whole network, we list all layers, inputs, outputs and kernel sizes in Table 3.2.

Table 3.2 The parameters of CNN architecture

Layer	Layer type	Kernel size	Stride	Output
Input			(120,120,32)	
C1	convolution	(1,5)	(1,1)	(120,120,32)
P1	max-pooling	(2,2)	(2,2)	(60,60,32)
C2	convolution	(1,5)	(1,1)	(60,60,64)
P2	max-pooling	(2,2)	(2,2)	(30,30,64)
C3	convolution	(1,5)	(1,1)	(30,30,128)
P3	max-pooling	(2,2)	(2,2)	(15,15,256)
C4	convolution	(1,3)	(1,1)	(15,15,256)
P4	max-pooling	(2,2)	(2,2)	(8,8,256)
F1	full-connected			1024
Output	softmax			3

3.6   LSTM for Solar Radio Spectrum Classification

All above models [2, 11, 12, 18] are static models, treating a spectrum as a spatial image. In fact, a spectrum describes the change of solar radio radiation at each frequency channel along with time. In [12], we made the first attempt to establish a multi-modal learning model, composed of an AE and a structured regularization term, where each frequency channel of a spectrum is regarded as a modality, so a spectrum is composed of multiple modalities. In low layers of the network, each modality was trained independently using an AE network. Meanwhile, all modalities were interacted through a structured regularization term. Then, two full-connected layers were stacked on the top of these AE layers. Finally, a softmax layer was stacked on the top of all hidden layers. Taking each column of a spectrum as one input, a spectrum then can be treated as a time series. Therefore, LSTM [19]–[20] was employed to further explore correlations among columns of a spectrum, resulting in better representation of spectrum, and correspondingly, better accuracy of spectrum classification.

The architecture of the proposed LSTM model for spectrum classification is illustrated in Fig. 3.9, where $(x_1, x_2, \cdots, x_T)$ is a time series consisting of the columns of a spectrum. As mentioned above, a spectrum can be treated as a time series. Thus, LSTM has advantage beyond static model to process spectrums.

A LSTM model consists of input layer, LSTM layer and softmax layer. In input layer, the vectors $(x_1, x_2, \cdots, x_T)$ are given by the columns of a spectrum. They are fed into the LSTM cells one by one in order. Each LSTM cell treats inputs according to (2.3)–(2.5). The recurrent concept of a LSTM network means that each LSTM cell does not output immediately responding to current input, while keeps silent until its time slot comes. The time slot is scheduled by time step which is given empirically. Then, a softmax layer is stacked upon the LSTM layer. It takes the LSTM output as the input, and outputs the spectrum type. In our case, the input is

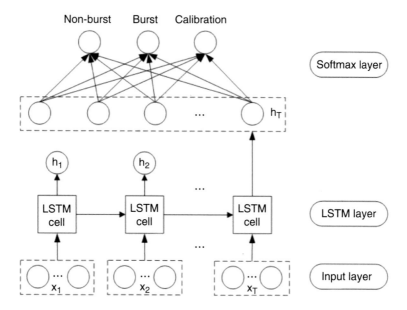

Fig. 3.9 Architecture of the proposed LSTM for spectrum classification

the size of 120×120. The time step is 120 so that each input spectrum has an output to the softmax layer.

LSTM has many variants, such as LSTM adding peephole connection [21], the Gated Recurrent Unit (GRU) [22]. In [23, 24], readers can find more about these LSTM variants. In this work, the initial one in [25] is used. The structure of a basic LSTM cell is illustrated in Fig. 2.4. In a LSTM cell, a memory block is provided to store the current status of the network so that this status can be kept for the next time step. By this way, the LSTM layer can explore interactions and correlations of a sequential input $x_1, x_2, \cdots, x_T$. Assumed $f_{LR}(\cdot)$ is the mapping function of a LSTM module, h_T represents the hidden state of a LSTM cell at the time T, the output of a LSTM layer is represented by

$$h_T = f_{LR}(x_1, x_2, \cdots, x_T), \tag{3.10}$$

where $\{x_i, i = 1, \ldots, T\}$ consists of a spectrum, the subscript i gives the time index. The LSTM unit f_{LR} is applied to $\{x_i, i = 1, \ldots, T\}$ to give an output h_T. Through (3.10), the highly compressed representation of a spectrum can be learnt. Afterwards, this representation is fed into a softmax layer, where a classifier is trained by supervised learning for classification. Assumed $\hat{L}$ is the output of the classifier, $\varphi[\cdot]$ is the function of classification, then the whole process of spectrum classification can be described by

$$\hat{L} = \varphi[f_{LR}(x_1, x_2, \cdots, x_T)] \tag{3.11}$$

For the task of spectrum classification, a softmax layer is stacked on the top of the LSTM layer, which is defined as

$$softmax(z_i) = \frac{\exp(z_i)}{\sum_j \exp(z_i)}. \tag{3.12}$$

The output of LSTM layer h_T firstly goes through a full connected network with 3 outputs as demonstrated in Fig. 3.9. This process can be described by

$$z = W_s h_T + b_s, \tag{3.13}$$

where W_s and b_s are the weights and bias of the fully connected layer connecting LSTM cell and softmax function. Then, z is fed into the softmax function (3.12) to output the probabilities of z belonging to the given three classes. This process is demonstrated in Fig. 3.10, where the input is a vector which dimension is dependent on the number of classes, the output must be less than 1 so that it can be the probability value.

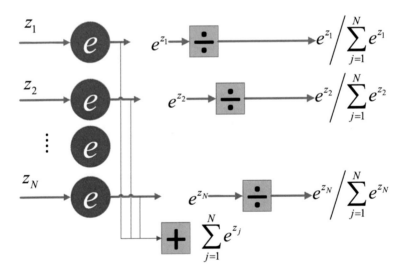

Fig. 3.10 Illustration of softmax layer for classification

3.7 Model Evaluation and Analysis

3.7.1 Database

The statistics indicate that the solar burst is very rare among all captured data. We have recorded millions of microwaves in total from 1995 to the end of 2001 by the SBRS [1]. However, only hundreds of them are "burst" as shown in Table 3.3.

With the aid of expertise of solar astronomy, 8816 spectrums were labelled to form a database. We firstly labeled spectrums into six categories (0: no burst or hard to identify, 1: weak burst, 2: moderate burst, 3: large burst, 4: data with interference, 5: calibration). Table 3.4 gives the number of spectrums for each type in the database. However, each category has insufficient samples, so we combine three burst types (1, 2, 3) into one category. Thus, there are three categories: "burst", "non-burst", and "calibration" in the database. Table 3.5 gives the number of spectrums of each category. A spectrum of "burst" at least contains a detectable solar radio burst during its lifetime, while a spectrum of "non-burst" has no identifiable burst during its lifetime. The "calibration" spectrum usually contains a step calibration signal. It is much easy to identified.

An original spectrum is of the size 120×2520. The horizontal axis is time, and the vertical axis is frequency. After pre-processing, new spectrum is of the size 120×120.

3.7.2 Evaluation Metrics

For evaluating a classifier, many evaluation indexes/metrics have provided, e.g., accuracy, error rate, precision and recall. They are defined on the basis of the following four basic terms.

Table 3.3 The statistics of solar radio bursts of SBRS

Freq. range(GHz)	0.5–1.5	1.0–2.0	2.6–3.8	4.5–7.5	5.2–7.6
Number of bursts	108	526	921	233	550

Table 3.4 The statistics of spectrum types ((0: no burst or hard to identify, 1: weak burst, 2: moderate burst, 3: large burst, 4: data with interference, 5: calibration))

Categories	0	1	2	3	4	Total
Number of spectrums	6670	618	268	272	570	8816

Table 3.5 The database of spectrums with three classes

Spectrum types	Non-burst	Burst	Calibration	Total
Number of spectrums	6670	1158	988	8816

(i) True positive (TP): if a positive instance is successfully predicted to be a position class;
(ii) False negative (FN): if a positive instance is wrong predicted to be a negative class;
(iii) False positive (FP): if a negative instance is predicted to be a positive class;
(iv) True negative (TN): if a negative instance is successfully predicted to be a negative class.

Over these four terms, the percentage/ratio of TP among all positive instances (TP+FN) is named true positive rate (TPR), i.e.,

$$TPR = TP/(TP+FN), \qquad (3.14)$$

representing how much percent of positive instances which are successfully retrieved (correctly classified into positive class) within all positive instances. Similarly, false positive rate (FPR) is defined as

$$FPR = FP/(FP+TN), \qquad (3.15)$$

indicating how much percent of negative instances which are wrong decided as positive instances within negative instance category. Let P and N represent the number of positive instance and negative instance, respectively. Accuracy index is defined as

$$Accuracy = (TP+TN)/(P+N), \qquad (3.16)$$

precision or precision ratio is defined as

$$Precision = TP/(TP+FP), \qquad (3.17)$$

recall or recall ration is defined as

$$Recall = TP/(TP+FN) = TP/P = sensitive. \qquad (3.18)$$

From definitions above, accuracy give the percentage of instances successfully classified in all instances (P+N is the total number of instances). The numerator is composed of all instances classified correctly, no matter positive instance or negative instance, while the denominator consists of all instances.

3.7.3 Model Evaluation

For evaluating a model, the database is split into two parts, training set and testing set. For training, 800 "burst", 800 "non-burst" and 800 "calibration" are randomly selected each time from the database to form a training set. A classifier is trained on this training set with the input of spectrums and their labels. The rest of database conforms the testing set. To efficiently assess the proposed LSTM model, we compare it with the previous DBN model [2], CNN model [18], multimodality model [11, 12]. It is also compared with a traditional model of PCA+SVM [2]. The codec and database used in this work can be accessed via https://github.com/filterbank/spectrumcls.

For fair comparison, we follow the evaluation indexes in the benchmarks [2, 11, 12, 18], TPR and FPR indexes as defined in (3.14) and (3.15) for evaluating the proposed LSTM model. The larger TPR and smaller FPR represents the better classifier.

In our case, we have three types of spectrums, so TPR and FPR are computed independently for each type. In addition, an interclass imbalance concerns our case, where there is only a small percentage of "burst" samples. For an interclass imbalance problem, the TRP and FPR indexes are better than the accuracy index for describing the performance of classification. There are two reasons for the short of "burst" samples. The one is that solar burst events are sparse among all recorded data. The other is that manually labelling is very difficult and time consuming.

After several rounds of training and testing, the statistics of TPR and FPR are listed in Table 3.6. Analyzing from Table 3.6, all of the five models on deep learning perform well on spectrum classification, while the method of PCA&SVM which was doable for general image fails on solar radio spectrum. For "calibration" type, all deep learning models achieve good performance. The TPR of all models is around 95%, which means 95% data is classified correctly. The FPR of all models is very small, less than 3.2%, which means only 3.2% of other types of data are wrongly classified into "calibration". For "burst" type, the proposed LSTM model achieves the highest TPR up to 85.4%, means 85.4% of "burst" data are correctly identified by the LSTM model. For "burst" type, the FPR of the LSMT model is 6.7%. It indicates that 6.7% of other types of spectrums are wrongly classified into "burst". From the definition of FPR, the smaller the FPR is given, the better the performance is achieved. Thus, the LSTM model is the best among all compared models regarding FPR. For "non-burst" type, the LSTM is the best among all compared models with respect to TPR. Comparing "burst" and "non-burst", both better TPR and FPR (larger TPR and smaller FPR) are for "non-burst". The reason is that "non-burst" has much simpler image pattern than "burst" as shown in Fig. 3.1. Thus, it is easier identified than "burst".

The gain of the proposed LSTM model may contribute to two aspects. First, a spectrum is reorganized into a time series so that its inner structure is exploited for classification. Second, the LSTM module can efficiently learn the relations and interactions of a time series to generate more representative features of a spectrum.

Table 3.6 Performance comparisons of solar radio spectrum classification

Category	Method											
	LSTM		CNN		Multi-modality Deep		Multi-modality		DBN		PCA+SVM	
	TPR(%)	FPR(%)	TPR(%)	FPR(%)	TPR(%)	FPR(%)	TPR(%)	FPR(%)	TPR(%)	FPR(%)	TPR(%)	FPR(%)
Burst	85.4	6.7	83.8	9.4	82.2	22.5	70.9	15.6	67.4	13.2	52.7	26.6
Non-burst	92.3	8.2	89.7	8.7	83.3	9.6	80.9	13.9	86.4	14.1	0.1	16.6
Calibration	96.2	0.9	100	0.7	92.5	1.7	96.8	3.2	95.7	0.4	38.3	72.2

References

1. Fu Q, Ji H, Qin Z, Xu Z, Xia Z, Wu H, Liu Y, Yan Y, Huang G, Chen Z et al. 2004 A new solar broadband radio spectrometer (SBRS) in China. *Solar Physics* **222**, 167–173.
2. Chen Z, Ma L, Xu L, Tan C, Yan Y. 2016 Imaging and representation learning of solar radio spectrums for classification. *Multimedia Tools and Applications* **75**, 2859–2875.
3. Hinton GE, Osindero S, Teh YW. 2006 A fast learning algorithm for deep belief nets. *Neural computation* **18**, 1527–1554.
4. Salakhutdinov R, Murray I. 2008 On the quantitative analysis of deep belief networks. In Cohen WW, McCallum A, Roweis ST, editors, *Machine Learning, Proceedings of the Twenty-Fifth International Conference (ICML 2008), Helsinki, Finland, June 5-9, 2008* vol. 307*ACM International Conference Proceeding Series* pp. 872–879. ACM.
5. Hinton GE, Salakhutdinov RR. 2006 Reducing the Dimensionality of Data with Neural Networks. *Science* **313**, 504–507.
6. Suykens JA, Vandewalle J. 1999 Least squares support vector machine classifiers. *Neural processing letters* **9**, 293–300.
7. Jolliffe I. 2011 Principal component analysis. In *International encyclopedia of statistical science* pp. 1094–1096. Springer.
8. Wold S, Esbensen K, Geladi P. 1987 Principal component analysis. *Chemometrics and intelligent laboratory systems* **2**, 37–52.
9. Vincent P, Larochelle H, Bengio Y, Manzagol PA. 2008 Extracting and Composing Robust Features with Denoising Autoencoders. In *Machine Learning, 25th International Conference on* pp. 1096–1103 New York, NY, USA. ACM.
10. Vincent P, Larochelle H, Lajoie I, Bengio Y, Manzagol PA. 2010 Stacked denoising autoencoders: Learning useful representations in a deep network with a local denoising criterion. *Journal of machine learning research* **11**, 3371–3408.
11. Chen Z, Ma L, Xu L, Weng Y, Yan Y. 2015 Multimodal learning for classification of solar radio spectrum. In *Systems, Man, and Cybernetics (SMC), 2015 IEEE International Conference on* pp. 1035–1040. IEEE.
12. Ma L, Chen Z, Xu L, Yan Y. 2017 Multimodal deep learning for solar radio burst classification. *Pattern Recognition* **61**, 573–582.
13. Vincent P, Larochelle H, Bengio Y, Manzagol PA. 2008 Extracting and composing robust features with denoising autoencoders. In *Proceedings of the 25th international conference on Machine learning* pp. 1096–1103. ACM.
14. Guillaumin M, Verbeek J, Schmid C. 2010 Multimodal semi-supervised learning for image classification. In *Computer Vision and Pattern Recognition (CVPR), 2010 IEEE Conference on* pp. 902–909. IEEE.
15. Ngiam J, Khosla A, Kim M, Nam J, Lee H, Ng AY. 2011 Multimodal deep learning. In *Proceedings of the 28th international conference on machine learning (ICML-11)* pp. 689–696.
16. Fawcett T. 2006 An introduction to ROC analysis. *Pattern recognition letters* **27**, 861–874.
17. Jalali A, Sanghavi S, Ruan C, Ravikumar P. 2010 A Dirty Model for Multi-task Learning. In Lafferty J, Williams C, Shawe-Taylor J, Zemel R, Culotta A, editors, *Advances in Neural Information Processing Systems* vol. 23. Curran Associates, Inc.
18. Chen S, Xu L, Ma L, Zhang W, Chen Z, Yan Y. 2017 Convolutional neural network for classification of solar radio spectrum. In *Multimedia & Expo Workshops (ICMEW), 2017 IEEE International Conference on* pp. 198–201. IEEE.
19. Hochreiter S, Schmidhuber J. 1997 Long short-term memory. *Neural computation* **9**, 1735–1780.
20. Gers FA, Schmidhuber J, Cummins F. 1999 Learning to forget: Continual prediction with LSTM.

21. Cho K, Van Merriënboer B, Gulcehre C, Bahdanau D, Bougares F, Schwenk H, Bengio Y. 2014 Learning phrase representations using RNN encoder-decoder for statistical machine translation. *arXiv preprint arXiv:1406.1078.*
22. Graves A, Jaitly N. 2014 Towards end-to-end speech recognition with recurrent neural networks. In *International Conference on Machine Learning* pp. 1764–1772.
23. Gers FA, Schraudolph NN, Schmidhuber J. 2002 Learning precise timing with LSTM recurrent networks. *Journal of machine learning research* **3**, 115–143.
24. Greff K, Srivastava RK, Koutník J, Steunebrink BR, Schmidhuber J. 2017 LSTM: A search space odyssey. *IEEE transactions on neural networks and learning systems* **28**, 2222–2232.
25. Jozefowicz R, Zaremba W, Sutskever I. 2015 An empirical exploration of recurrent network architectures. In *International Conference on Machine Learning* pp. 2342–2350.

Chapter 4
Deep Learning in Solar Object Detection Tasks

Abstract Solar observation provides us abundant solar images containing plentiful information about solar activities. Especially, solar instruments onboard satellite continuously record high-resolution and high-cadence full-disk solar images. These images are used for solar activity forecasting and statistical analysis. Usually, it is required to mine key information from full-disk images firstly. Then, over extracted information, one can establish classification, recognition or forecasting models by using machine learning or deep learning. In a full-disk solar image, active region, filament, coronal hole and sunspot are the objects carrying major information about solar activities. In computer vision, object detection is one of the most classical tasks, which has been well investigated. In this chapter, we present two examples of object detection from solar image, by using well pre-trained deep learning models in computer vision.

Keywords Active region (AR) · Object detection · Regional convolutional neural network (R-CNN) · Region proposal network (RPN)

4.1 Active Region (AR) Detection

Solar eruptive events would affect the radio communication, Global Positioning Systems, and some high-tech equipments in both space and ground. Active regions on the Sun are the main source regions of solar eruptive events. Therefore, the automatic detection of active regions is important not only for the routine observation, but also for the solar activity forecast. At present, active regions are manually extracted or automatically extracted by using traditional image processing techniques. Since active regions constantly evolve, it's not easy to design a suitable feature extractor. In this work, two representative object detection models, Faster R-CNN and YOLO V3, are employed to learn the characteristics of active regions, and then establish active region detection models. In addition, pre-trained models of Faster R-CNN and YOLO V3 are used, and refined over full-disk solar magnetograms. The experimental results show that both two models achieve high accuracy of active region detection. In addition, YOLO V3 is 4% and 1% better than Faster R-CNN in terms of True Positive (TP) and True Negative (TN) indexes, respectively, meanwhile the YOLO V3 is 8 times faster than Faster R-CNN.

4.1.1 Active Region (AR)

A solar active region is an area with strong magnetic field on the Sun. It is considered as the major source region of solar eruptive events. The region frequently breeds various types of solar activity, including explosive solar bursts, such as solar flares and coronal mass ejections (CME). Active region has strong magnetic field which is usually strong enough to inhibit convection, preventing convection transporting energy from the Sun's interior to the photosphere. Thus, the temperature of active region decreases relative to its surrounding, causing cooler plasma which is known as sunspots. Sunspot is a visual indicator of active region. High-energy phenomena associated with active region makes this region bright in ultraviolet and X-ray image of the Sun. Many types of dramatic solar features, such as solar prominences and coronal loops, frequently appear around active region. In Fig. 4.1, an example of active region is demonstrated, including AIA 304 Å, AIA 193 Å, AIA 304 Å, AIA 1700 Å, HMI Magnetogram, HMI 6173 Å and XRT. Figure 4.1a, c, and e

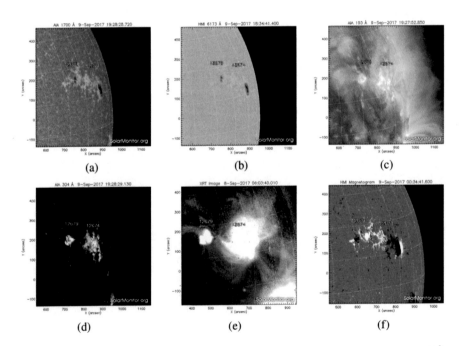

Fig. 4.1 Big solar flare occurring in Sept. 9, 2017 around active regions 12674, 12679. (**a**) 1700Å. (**b**) HMI 6173Å. (**c**) 193Å. (**d**) 304Å. (**e**) HXR. (**f**) HMI Magnetogram

demonstrates obvious active region, Fig. 4.1b show an ultraviolet view of coronal loops in the Sun's atmosphere above active region. Figure 4.1d shows an intensive brightening on X-ray image. The solar eruptive events could cause severe space weather effects, which may affect the safety of satellites, the precision of Global Positioning Systems and so on. Therefore, it is of great importance in the routine monitoring and automatic extraction of active regions.

4.1.2 State-of-the-Art of Active Region Detection

Some efforts have been made towards automatically identifying solar active regions. Benkhalil et al. [1] determined thresholds for active region to get initial seeds of active regions. The noise is removed by median filtering and morphological operations. Based on these initial seeds, region growing algorithm is used to detect active regions. Zhang et al. [2] designed an active region detection system by using intensity threshold and morphological analysis algorithm. McAteer et al. [3] combined region growing algorithm and boundary extraction technique to detect active regions. Caballero and Aranda [4] proposed a two-step method to detect active regions from full-disk images. In the first step, region growing algorithm is applied to segment the bright parts in active regions. In the second step, partition based clustering and hierarchical clustering are used to group together these bright parts, respectively. The hierarchical clustering method was recommended because of its good performance. Higgins et al. [5] proposed the Solar Monitor Active Region Tracking (SMART) algorithm to detect and track active regions throughout their life time. In this algorithm, quiet Sun and some transient magnetic features were removed, and then the region-growing technique were applied to determine active regions. Some magnetic properties of active region such as region size, magnetic flux emergence rate, non-potentiality measurements are calculated. Colak and Qahwaji [6] proposed the Automated Solar Activity Prediction tool (ASAP) to automatically detect, group and classify sunspots. The intensity threshold, morphological algorithms, region growing algorithms and neural networks are applied to determine the boundaries of sunspots. Watson et al. [7] proposed the Sunspot Tracking And Recognition Algorithm (STARA) to detect and track sunspots from solar white light images. Barra et al. [8] proposed a fuzzy clustering algorithm (spatial possibilistic clustering algorithm (SPoCA)) to automatically segment the full-disk solar images into Coronal Holes, Quiet Sun and Active Regions, respectively. The SPoCA algorithm was improved in [9], and automatic tracking of active regions was further developed. The performances of SMART, ASAP, STARA and SPoCA were analyzed and compared in [10]. They found that ASAP tends to detect very small sunspots while STARA has a higher threshold for sunspot detection, and SMART and SPoCA detect more regions than the National Oceanic and Atmospheric Administration (NOAA). The proposed detection methods mainly based on intensity threshold, morphological operations, region growing algorithms and clustering methods. In these methods, pre-defined

Table 4.1 Related work

Related work	Algorithms				
	Threshold	Morphology	Region growing	Clustering	Deep learning
[1]	✓		✓		
[2]	✓	✓			
[3]			✓		
[4]			✓	✓	
[5]			✓		
[6]	✓	✓	✓		
[8]				✓	
Our					✓

parameters should be determined [11]. However, it is not easy to set optimal parameters.

Deep learning algorithm can automatically extract the distinguishing features and realize end-to-end objective detection. So far, deep learning algorithm has not been applied to detect solar active regions. Here, faster R-CNN (Regions with Convolutional Neural Networks) algorithms are used to detect active regions from full-disk solar images, and it's performance is compared with the NOAA labeled active regions.

There have been many works on automatic detection of active regions, including intensity threshold, morphological operations, region growing algorithms, clustering methods and the combination of them shown in Table 4.1. However, it is difficult to determine parameters in these algorithms.

Deep learning method has been proposed to learn a detection model from the observational data, including R-CNN, Fast R-CNN, Faster R-CNN and more new algorithms. There are three steps to generate a R-CNN detector. In first step, region proposals are generated. And then, a CNN is trained to classify these proposal regions. Finally, the bounding boxes of the proposal regions are refined. Unlike the R-CNN algorithm, fast R-CNN algorithm maps CNN features corresponding to generated region proposals, hence, the fast R-CNN detector is more efficient than the R-CNN detector. In Faster R-CNN detector, region proposal network (RPN) is used to generate region proposals. RPN is faster and better than the proposed generation method of region proposals. In this paper, we implement a deep learning based active region detection model for the first time.

4.1.3 *Deep Learning Based Object Detection*

Object detection is one of fundamental problem in computer vision. Before deep learning, it already has been widely investigated. In [12], a histogram of oriented gradients (HOG) detector was raised under the principle that the appearance and shape of the target object can be well described by the directional density distribution of the gradient or edge in an image. In addition, it is designed to be computed on a dense grid of uniformly spaced cells and use overlapping local contrast normalization (on "blocks") for handling the feature invariance (including translation, scale, illumination, etc.) and the nonlinearity (on discriminating different objects categories). In [13], a deformable part-based model (DPM) was raised for solving the multi-scale problem in HOG, which was the winners of VOC-07, 08 and 09 detection challenges. A typical DPM detector consists of a root-filter and a number of part-filters. Instead of manually specifying the configurations of the part filters (e.g., size and location), a weakly supervised learning method is developed in DPM where all configurations of part filters can be learned automatically as latent variables. As a method of hand-crafted features, DPM cannot go beyond deep learning methods characterized by automatic feature learning with respect to detection accuracy and computing complexity although there were many tricks for augmenting DPM.

The earliest deep learning based algorithm was a two-stage detector, namely R-CNN (regional convolutional neural network) [14]. In Fig. 4.2, the architecture of R-CNN is illustrated. It starts with the extraction of a set of object proposals by selective search, around 2000 bottom-up proposals. Then, these proposals are rescaled to a uniform size, and fed into a CNN model pre-trained on ImageNet to extract features. Finally, a SVM classifier is stacked on the top of feature extractor to predict the presence of an object within each region and to recognize object categories. R-CNN achieved a significant performance boost on VOC07, with a large improvement of mean average precision (mAP) from 33.7% (DPM-v5) to 58.5%. Its drawback lies in the redundant feature computations on a large number of overlapped proposals, leading to an extremely slow detection speed. SPPNet [15] was proposed to overcome this problem, which enables a CNN to generate a fixed-length representation regardless of the size of image/region of interest without rescaling it. In SPPNet, feature maps are computed from the entire image only once, and then fixed length representations of arbitrary regions can be generated for training the detectors, which avoids repeatedly computing the convolutional features.

By computing feature maps of an entire image only once, SPPNet achieved more than 20 times faster than R-CNN without sacrificing any detection accuracy.

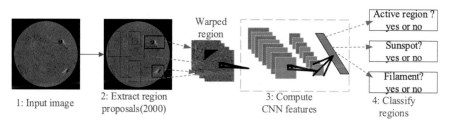

Fig. 4.2 Flowchart of R-CNN

However, it is still multi-stage, and only fine-tuning its fully connected layers while simply ignoring all previous layers. Thus, Fast R-CNN [16] was later proposed for solving these problems. It enables us to simultaneously train a detector and a bounding box regressor under the same network configurations. In Fast R-CNN, RoI layer extracts multiple regions of interest (RoIs) and sent the feature into a fully convolutional network. Each RoI is pooled into a fixed-size feature map and then mapped to a feature vector by fully connected layers (FCs). The network has two output vectors per RoI, softmax probabilities and per-class bounding-box regression offsets. The most remarkable feature of Fast R-CNN is an end-to-end architecture, which integrating the advantages of R-CNN and SPPNet, increasing the mAP from 58.5% (R-CNN) to 70.0% over VOC07 dataset, accelerating detection speed over 200 times faster than R-CNN. However, its detection speed is still limited by the proposal detection until the proposal of Faster R-CNN [17]. Faster R-CNN was towards real-rime object detection with a region proposal network (RPN) which takes feature maps as input and outputs a set of rectangular object proposals each of which has an objectness score and a bounding box.

The main contribution of Faster R-CNN is the introduction of RPN that enables nearly cost-free region proposals. From R-CNN to Faster R-CNN, most individual blocks of an object detection system, e.g., proposal detection, feature extraction, bounding box regression, etc., have been gradually integrated into a unified, end-to-end learning framework.

To further reduce computation redundancy at subsequent detection stage, a variety of improvements have been proposed, such as RFCN [18] and Light head R-CNN [19].

The series of R-CNNs is typical a two-stage algorithm, consisting of proposals and classification/regression stages. The algorithm of YOLO [20] is an one-stage algorithm. It splits the input image into an $S \times S$ grid. If the center of an object falls into a grid cell, that grid cell is responsible for detecting that object. Each grid cell predicts B bounding boxes and confidence scores for those boxes. These confidence scores reflect how confident the model is that the box contains an object and also how accurate it thinks the box is that it predicts. Each bounding box consists of 5 predictions, x, y, w, h, and confidence. The (x, y) coordinates represent the center of the box relative to the bounds of the grid cell. The width and height are predicted relative to the whole image. Finally, the confidence prediction represents the IoU between the predicted box and any ground truth box. YOLO [20] is the first one-stage detector in deep learning era. YOLO has greatly improved detection speed, However, it suffers from a drop of the localization accuracy compared with two-stage detectors, especially for some small objects. YOLO's subsequent versions [21, 22] and the latter proposed SSD [23] has paid more attention to this problem.

The SSD approach is based on a feed-forward convolutional network that produces a fixed-size collection of bounding boxes and scores for the presence of object class instances in those boxes, followed by a non-maximum suppression step to produce the final detections. The model loss is a weighted sum between localization loss (e.g. Smooth L1) and confidence loss (e.g. Softmax). The main contribution of SSD is the introduction of the multi-reference and multi-resolution detection techniques, which significantly improves the detection accuracy of a one-stage detector, especially for some small objects.

In one-stage algorithm, the extreme foreground-background class imbalance would be encountered during training of dense detectors. To this end, a new loss function named "focal loss" has been introduced in RetinaNet [24] by reshaping the standard cross entropy loss so that detector will highlight on hard, misclassified examples during training. RetinaNet is a single, unified network composed of a backbone network and two task-specific subnetworks. The backbone is responsible for computing a convolutional feature map over an entire input image and is an off-the-self convolutional network. The first subnet performs convolutional object classification on the backbone's output; the second subnet performs convolutional bounding box regression. The network design is intentionally simple, which enables this work to focus on a novel focal loss function that eliminates the accuracy gap between our one-stage detector and state-of-the-art two-stage detectors like Faster R-CNN with RPN while running at faster speeds. Focal Loss enables the one-stage detectors to achieve comparable accuracy of two-stage detectors while maintaining very high detection speed.

4.1.4 Our Proposal of Active Region Detection

Traditional object detection techniques include 3 major steps:

(1) Region proposal generation. A large number of region proposals are generated by Selective Search algorithm [25].
(2) Feature extraction. Some feature extractors are applied to get a fixed-length feature vector, for example, Hog or SIFT [26, 27].
(3) Classification. Based on the fixed-length feature vector, classification model can be learned to judge whether the region proposal is the object.

The feature extraction is the key of the success of object detection. In traditional object detection methods, it's difficult to design a feature extractor for a specific task. In deep learning based methods, convolutional neural network (CNN) is used to extract image features. For example, region-based convolutional neural network (R-CNN) uses selective search to prepare regional proposals. Then, scaling all proposals to the same size, CNN is employed to extract image features of proposals, followed by a classifier and a regressor to classify object and background, and compute coordinates of an object respectively. To further speed up object detection, Fast R-CNN was further optimized to get an end-to-end network, namely Faster R-CNN, by employing region proposal network (RPN) to replace selective search to generate region proposals, saving computing time dramatically.

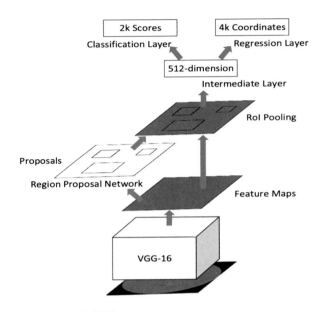

Fig. 4.3 Flowchart of Faster R-CNN

As shown in Fig. 4.3, Faster R-CNN is composed of 3 components [28]:

(1) A pre-trained CNN (VGG-16 [29]) is used to extract image features from solar full-disk magnetograms;
(2) RPN is performed over feature maps derived from the first step to generate object proposals;
(3) A classifier and a regressor are trained to classify object and background, and derive coordinates of each object.

The feature extractor can be a pre-trained CNN, e.g., VGG-16 which includes 16 layers of convolution layers and pooling layers. Convolution layers are responsible for feature extraction, while pooling layers are used to reduce feature dimension. In our implementation, full-disk image is first resized to 1024×1024. An image of dimension 1024×1024 is then reduced to a feature map of 64×64 through feature extractor. Furthermore, a RPN is trained to generate object proposals to replace time consuming selective search module in R-CNN and Fast-R-CNN.

To train a RPN, labelled samples are needed. The positive class means the concerned anchor contains an object, while the negative class means the concerned anchor is the background. Intersection over Union (IoU) is defined to be the ratio of area of overlapped region between object proposal and ground truth relative to current whole union. The proposal object which has an IoU overlap higher than 0.7 with a ground-truth box is considered as a positive sample, while the proposal object which has an IoU overlap less than 0.3 with a ground-truth box is considered as a negative sample. To augment dataset, random transforming is used to generate more samples to reduce the risk of over-fitting. After getting regional proposals through

RPN, the RoI pooling layer is implemented to get uniform-sized regional proposals. And then, the softmax classifier and the bounding box regressor are trained to determine the class of regional proposal (object or background) and its bounding box.

4.1.5 Performance Evaluation

The Helioseismic and Magnetic Imager (HMI) onboard the Solar Dynamics Observatory (SDO) provides the routine full-disk magnetic observation of the Sun. The National Oceanic and Atmospheric Administration (NOAA) maintains an active region list. It detects active regions manually or automatically everyday, and gives each of them in chronological order.

We got solar full-disk magnetograms and corresponding active region summary from the Joint Science Operations Center (JSOC)[1] and Space Weather Prediction Center of National Oceanic (SWPC) and Atmospheric Administration (NOAA).[2]

There are 4645 full-disk magnetograms with time interval of 24 h from 2010 to 2017. A database of active region is established over these magnetograms and corresponding active region information. This database is split into training set and testing set, the former contains 4073 samples from 2010 to 2015, while the later includes remaining samples from 2016 to 2017.

The proposed solar active region detection model is optimized using the SGD algorithm with *momentum* = 0.9, batch size of 4, maximum epoch of 100. The initial learning rate is 0.001 and then is divided by 10 after every 6 epochs.

Figure 4.4 shows an example of successful active region detection, where 4 active regions (recorded on Aug. 26, 2016) are successfully detected. However, our model may miss some active regions sometimes, as shown in Fig. 4.5 where a magnetogram was recorded on Mar. 30, 2017. The TP rate is 90%, which means that 10% active regions would be missed. Usually, the small active regions or the active regions around the edge of solar disk are more likely to be missed. Because active regions around solar disk edge are influenced by the projection effect of the Sun. Besides, our model may falsely judge some quiet regions as active regions, as shown in Fig. 4.6 where a magnetogram was recorded in May 4, 2016. The TN rate is 98%, which means that only 2% quiet regions could be falsely detected.

---

[1] http://jsoc.stanford.edu/ajax/lookdata.html.

[2] https://www.swpc.noaa.gov/products/solar-region-summary.

Fig. 4.4 Successful active
region detection

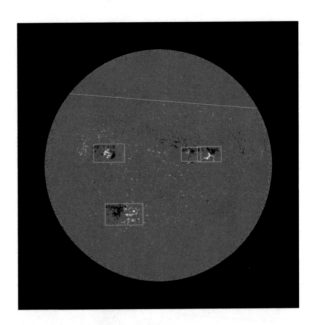

Fig. 4.5 Missing active
region

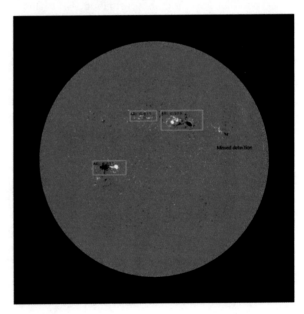

Fig. 4.6 False detection of active region

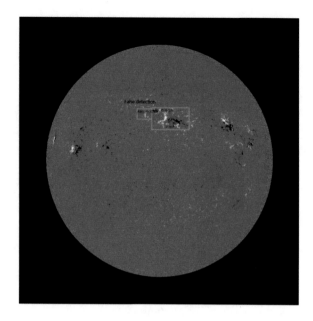

4.1.6   Conclusions

An active region detection dataset is build over SDO/HMI magnetograms from 2010 to 2017. The dataset consists of solar full-disk magnetograms, active region and quiet region proposals which are positive and negative samples respectively, and bounding boxes of active regions. Over the established dataset, a Faster R-CNN model for active region detection is trained. The performance evaluation indicates that 10% active regions are missed, while, 2% quiet regions are falsely detected as active regions, demonstrating good performance of our proposed active region detection model. However, there are still some problems concerning small active region or active region around the edge of solar disk. Further optimization could be done by specially treating small active regions and active regions influenced by the projection effect around the edge of solar disk.

4.2   EUV Waves Detection

4.2.1   EUV Waves

Coronal "EUV waves" appear as EUV bright fronts propagating across a significant part of solar disk. For example, [30] once reported a large-scale wave, with bright fronts followed by extending dimmings, and propagating out from a flaring site. Although this phenomenon has been widely studied with the high spatio-temporal observation of SOHO/EIT and SDO/AIA, its interpretation (or physical principle), is still uncertain in the community. One branch still follows the wave interpretation, treating EUV waves as slow-mode waves [31], slow-mode solitons (e.g., [32]), or shock waves more generally. For more details, one can consult the review paper [33]. The other branch no longer holds the view of waves, treating EUV waves as pseudo-waves which are caused by coronal magnetic-field reconfiguration during corona mass ejection (CME), where EUV wave brightenings possibly result from stretching of magnetic-field lines [34], Joule heating in a current shell [35], or continuous small-scale reconnection [36].

There has been a growth of strong evidence to suggest that EUV waves are closely related to CMEs, rather than flares [37, 38]. In addition, CMEs are preceded by EUV waves. Thus, by detecting EUV waves as they traverse solar disk, we can early knows CMEs before coronagraph. In [39], we investigated EUV waves detection from SDO/AIA image by using deep learning. In addition, a model was trained to extract outlines of EUV waves, namely wavefronts of EUV waves. With outlines of EUV waves, attributes (physical and topological parameters) of CMEs, like initial/final velocity, accelerated velocity, area and angular can be deduced easily. Further, CME can be early predicted before coronagraph. The proposed model consists of a detection network to detect EUV waves and a GAN network to outline wavefronts of EUV waves.

4.2.2 Detection Network for EUV Waves

The existing classical detection networks include R-CNN [14], Fast R-CNN [16], Faster R-CNN [17] and Mask-RCNN [40]. They treat object detection as a classification task. For detecting an object, many candidate region proposals are firstly catered for checking. These region proposals are extracted from different position and with different patch size. Then, a binary classifier is trained on the correspondences between region proposals and ground-truth, where the correspondence indicates whether a region proposal is the target. The region proposal with the best matching between itself and ground-truth is the target. Compared to the former networks [14–16], Faster R-CNN is more computationally efficient due to the usage of Region Proposal Networks(RPN) [17], so it is chosen for building our EUV wave detection network.

The proposed detection network is illustrated in Fig. 4.7. It consists of a convolution module for extracting features, a RPN module for determining the best region proposal, a Region of Interest (RoI) pooling [15] module, a classifier and a regressor. To generate region proposals, we slide a small network over the final convolutional feature map. This network is fully connected to an $n \times n$ spatial window of the input convolutional feature map. Each window is converted to a vector, and fed into two sibling fully-connected layers, namely a box-regression layer (denoted by "reg") and a box-classification layer (denoted by "cls"). At each sliding window location, we extract k region proposals with different scales and

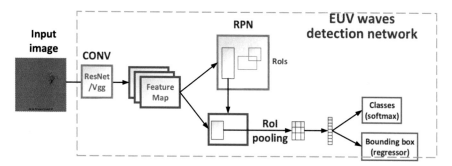

Fig. 4.7 EUV waves detection network

aspect ratio. The "reg" can locate each region proposal, outputting its top-left and lower-right coordinates, while "cls" discriminates the foreground from background. Thus, the "reg" and "cls" have $4k$ and $2k$ outputs, respectively, as shown in Fig. 3.9. Each region proposal is named an anchor. Here, we use 5 scales and 3 aspect ratios, yielding $k = 15$ anchors at each sliding position. For a $W \times H$ convolutional feature map, there are $W \times H \times k$ anchors in total.

Given an image, we firstly use a pre-trained network (CONV), e.g., VGGNet [41] and ResNet [29], to extract image features. The pre-trained VGGNet and ResNet were provided for downloading via GitHub, e.g., [42]. Then, the RPN is stacked on the CONV to generate region proposals. In both R-CNN [14] and Fast R-CNN [16], selective search algorithm [25] is used to generate region proposals. It is implemented by CPU outside of neural network, resulting in very high computational complexity. As the RPN illustrated in Fig. 4.8 is used instead of selective search algorithm in Faster R-CNN, region proposals are generated in an end-to-end system. In addition, they share the same feature maps with the detection network, without overhead computing. Thus, Faster R-CNN is 10 times faster than Fast R-CNN for inference.

For training the RPN, a binary class label (of being a CME or not) is assigned to each anchor. If an anchor has an Intersection-over-Union (IoU) overlap higher than 0.7 relative to any one of the ground-truths, a positive label is assigned. While a negative label is assigned to an anchor whose IoU overlap is lower than 0.3 relative to all ground-truths. Then we can get our loss function for an image

$$\mathcal{L}(p_i, t_i) = \frac{1}{N_{\text{cls}}} \sum_i \mathcal{L}_{\text{cls}}(p_i, p_i^*) + \lambda \frac{1}{N_{\text{reg}}} \sum_i p_i^* \mathcal{L}_{\text{reg}}(t_i, t_i^*), \qquad (4.1)$$

where i is an index of an anchor in a mini-batch and p_i is the predicted probability of the i-th anchor being an object (CME). The ground-truth label p_i^* expresses the anchor is positive ($p_i^* = 1$) or negative ($p_i^* = 0$). t_i is the coordinates of the predicted bounding box, t_i^* is that of the ground-truth box associated with a positive anchor. $\mathcal{L}_{\text{cls}}$ is log loss over two classes (object and non-object). $\mathcal{L}_{\text{reg}}$

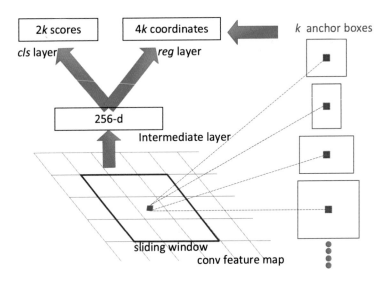

Fig. 4.8 Region Proposal Network (RPN)

constrains the predicted boxes's coordinates close to the ground-truth boxes. N_{cls} and N_{reg} are normalization factors which set to mini-batch size and the number of anchor locations. λ is set to 10, which will be further discussed in the following experiments.

Observing Fig. 3.9a, after region proposals are generated through the RPN, they go through the RoI pooling so that each region proposal has the same dimension. At last, these vectors of same dimension are fed into a classifier and a regressor, which output two vectors per RoI, softmax probabilities and per-class bounding-box regression offsets. We follow the Faster R-CNN, using a multi-task loss consisting of classification and regression.

4.2.3 Network of EUV Wave Outlining

After detecting EUV waves given by a rectangular region, we further draw outlines of EUV waves for better identifying its origin, propagation direction/speed and area. For this purpose, we employe pix2pix GAN to generate outlines of EUV waves automatically. Pix2pix GAN is a conditional GAN, consisting of a generator (G) and a discriminator (D). The generator is trained to forge "fakes" that cannot be distinguished from "real" by the discriminator. This training procedure is diagrammed in Fig. 4.9, where x, $G(x)$ and y are input image, generated image ("fake") and labeled image ("real"), respectively. For training model, we manually labeled EUV waves ("y") as demonstrated in Fig. 4.9.

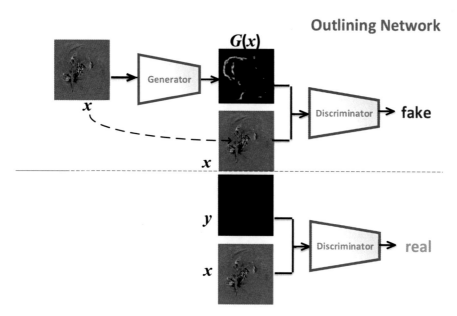

Fig. 4.9 Outlining wavefronts of detected EUV waves using pix2pix GAN

For training EUV waves outlining network, the following three processes are implemented iteratively until satisfying end condition or maximum iteration. Here, "x" represents an image patch containing EUV wave detected from the above detection network, "y" is the ground-truth which was drawn manually.

(i) Input an image patch (detected EUV wave in the first step) to G, and get an output $G(x)$;
(ii) Use $(x, G(x))$: "fake" and $\{x, y\}$: "real" pairs to train the discriminator D;
(iii) Use $(x, G(x))$ to train the generator G while fixing the discriminator D, and return loss;
(iv) Go to step (i).

In training stage, the goal of G is to gorge images as real as possible relative to ground-truth, so that D can be deceived. D tries its best to distinguish the forged $G(x)$ from "real" ones. Thus, G and D composes a two-player zero-sum game as described in (2.7). For our task, G is required to generate images with not only contour style but also close to ground-truth. Thus, beyond traditional cGAN, loss function of our model is defined as

$$G^* = \arg \min_G \max_D \mathcal{L}_{cGAN}(G, D) + \mathcal{L}_{L1}(G), \qquad (4.2)$$

where $\mathcal{L}_{L1}(G) = \mathbb{E}_{x,y,z}[\|y - G(x, z)\|_1]$.

After above two steps, we can finally describe EUV waves by simple outlines. Providing the sequence of outlines, we can more easily make inference about evolution/propagation of the EUV wave. Then, we can deduce initial/final velocity, accelerated velocity, area and main angular of a CME around its origin, characterizing a CME as the CDAW SOHO/LASCO CME catalog [43, 44].

4.2.4 Performance Evaluation

For probing origin of a CME and diagnosing its source-region, our studies were performed on SDO/AIA observation as CMEs preceded by EUV waves traverse solar disk. The testing conditions: Python, PyTorch deep learning package, Nvidia Geforch GTX780 GPU card. The databases and codes used in this work can be accessed via https://github.com/filterbank/EUVwaves.

We gathered 29 video sequences of X-ray flares from March 2011 to March 2015. For each video sequence, difference of each frame relative to the first frame was computed, resulting in better exhibition of a flare. First, a Faster R-CNN was trained for EUV waves detection, where ground-truth was provided by a bounding box deduced from the coordinates of active region. The stochastic gradient descent with the momentum of 0.9 was employed, with initial learning rate of 0.001, and declining by 0.1 after about 6 epochs. After EUV waves detection, we labelled wavefronts of EUV waves manually as ground-truth for training EUV waves outlining network. A pix2pix GAN was taken for EUV waves outlining as illustrated in Fig. 4.9, where "y" represents ground-truth, "x" denotes EUV waves detected by the Faster R-CNN in the first step.

After above two steps, outlines of EUV waves can be obtained successfully, as illustrated in Fig. 4.10, which can describe a CME around its origin more clearly. In addition, automatic outlining EUV waves can provide contours of EUV waves, which makes computing parameters of a CME easier.

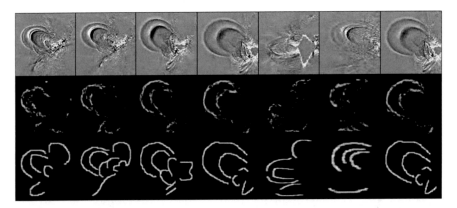

Fig. 4.10 EUV wave outlines generated by the proposed EUV wave outlining network (The first row shows EUV wave detection results; the middle row gives EUV wave outlines generated from the first row by the proposed EUV wave outlining model, and the last row lists ground-truth labelled manually)

References

1. Benkhalil A, Zharkova V, Zharkov S, Ipson S. 2006 Active region detection and verification with the solar feature catalogue. *Solar Physics* **235**, 87–106.
2. Zhang J, Wang Y, Liu Y. 2010 Statistical properties of solar active regions obtained from an automatic detection system and the computational biases. *The Astrophysical Journal* **723**, 1006.
3. McAteer RJ, Gallagher PT, Ireland J, Young CA. 2005 Automated boundary-extraction and region-growing techniques applied to solar magnetograms. *Solar Physics* **228**, 55–66.
4. Caballero C, Aranda M. 2013 A comparative study of clustering methods for active region detection in solar EUV images. *Solar Physics* **283**, 691–717.
5. Higgins PA, Gallagher PT, McAteer RJ, Bloomfield DS. 2011 Solar magnetic feature detection and tracking for space weather monitoring. *Advances in Space Research* **47**, 2105–2117.
6. Colak T, Qahwaji R. 2009 Automated solar activity prediction: a hybrid computer platform using machine learning and solar imaging for automated prediction of solar flares. *Space Weather* **7**.
7. Watson F, Fletcher L, Dalla S, Marshall S. 2009 Modelling the longitudinal asymmetry in sunspot emergence: the role of the Wilson depression. *Solar Physics* **260**, 5–19.
8. Barra V, Delouille V, Hochedez JF. 2008 Segmentation of extreme ultraviolet solar images via multichannel fuzzy clustering. *Advances in Space Research* **42**, 917–925.
9. Barra V, Delouille V, Kretzschmar M, Hochedez JF. 2009 Fast and robust segmentation of solar EUV images: algorithm and results for solar cycle 23. *Astronomy & Astrophysics* **505**, 361–371.
10. Verbeeck C, Higgins PA, Colak T, Watson FT, Delouille V, Mampaey B, Qahwaji R. 2013 A multi-wavelength analysis of active regions and sunspots by comparison of automatic detection algorithms. *Solar Physics* **283**, 67–95.
11. Harker BJ. 2012 Parameter-free automatic solar active region detection by Hermite function decomposition. *The Astrophysical Journal Supplement Series* **203**, 7.
12. Dalal N, Triggs B. 2005 Histograms of Oriented Gradients for Human Detection. In *2005 IEEE Computer Society Conference on Computer Vision and Pattern Recognition (CVPR 2005), 20-26 June 2005, San Diego, CA, USA* pp. 886–893. IEEE Computer Society.
13. Felzenszwalb PF, McAllester DA, Ramanan D. 2008 A discriminatively trained, multiscale, deformable part model. In *2008 IEEE Computer Society Conference on Computer Vision and Pattern Recognition (CVPR 2008), 24-26 June 2008, Anchorage, Alaska, USA*. IEEE Computer Society.
14. Girshick R, Donahue J, Darrell T, Malik J. 2014 Rich feature hierarchies for accurate object detection and semantic segmentation. In *Proceedings of the IEEE conference on computer vision and pattern recognition* pp. 580–587.
15. He K, Zhang X, Ren S, Sun J. 2015 Spatial pyramid pooling in deep convolutional networks for visual recognition. *IEEE transactions on pattern analysis and machine intelligence* **37**, 1904–1916.
16. Girshick R. 2015 Fast r-cnn. In *Proceedings of the IEEE international conference on computer vision* pp. 1440–1448.
17. Ren S, He K, Girshick R, Sun J. 2015 Faster r-cnn: Towards real-time object detection with region proposal networks. In *Advances in neural information processing systems* pp. 91–99.
18. Dai J, Li Y, He K, Sun J. 2016 R-FCN: Object Detection via Region-based Fully Convolutional Networks. In Lee DD, Sugiyama M, von Luxburg U, Guyon I, Garnett R, editors, *Advances in Neural Information Processing Systems 29: Annual Conference on Neural Information Processing Systems 2016, December 5-10, 2016, Barcelona, Spain* pp. 379–387.
19. Li Z, Peng C, Yu G, Zhang X, Deng Y, Sun J. 2017 Light-Head R-CNN: In Defense of Two-Stage Object Detector. *CoRR* **abs/1711.07264**.
20. Redmon J, Divvala SK, Girshick RB, Farhadi A. 2016 You Only Look Once: Unified, Real-Time Object Detection. In *2016 IEEE Conference on Computer Vision and Pattern Recognition, CVPR 2016, Las Vegas, NV, USA, June 27-30, 2016* pp. 779–788. IEEE Computer Society.

21. Redmon J, Farhadi A. 2017 YOLO9000: Better, Faster, Stronger. In *2017 IEEE Conference on Computer Vision and Pattern Recognition, CVPR 2017, Honolulu, HI, USA, July 21-26, 2017* pp. 6517–6525. IEEE Computer Society.
22. Redmon J, Farhadi A. 2018 YOLOv3: An Incremental Improvement. *CoRR* **abs/1804.02767**.
23. Liu W, Anguelov D, Erhan D, Szegedy C, Reed SE, Fu C, Berg AC. 2015 SSD: Single Shot MultiBox Detector. *CoRR* **abs/1512.02325**.
24. Lin T, Goyal P, Girshick RB, He K, Dollár P. 2017 Focal Loss for Dense Object Detection. *CoRR* **abs/1708.02002**.
25. Uijlings JR, Van De Sande KE, Gevers T, Smeulders AW. 2013 Selective search for object recognition. *International journal of computer vision* **104**, 154–171.
26. Lowe DG. 2004 Distinctive image features from scale-invariant keypoints. *International journal of computer vision* **60**, 91–110.
27. Dalal N, Triggs B. 2005 Histograms of oriented gradients for human detection. In *2005 IEEE computer society conference on computer vision and pattern recognition (CVPR'05)* vol. 1 pp. 886–893. Ieee.
28. Ren S, He K, Girshick R, Sun J. 2016 Faster R-CNN: towards real-time object detection with region proposal networks. *IEEE transactions on pattern analysis and machine intelligence* **39**, 1137–1149.
29. He K, Zhang X, Ren S, Sun J. 2016 Deep residual learning for image recognition. In *Proceedings of the IEEE conference on computer vision and pattern recognition* pp. 770–778.
30. Thompson BJ, Plunkett SP, Gurman JB, Newmark JS, St. Cyr OC, Michels DJ. 1998 SOHO/EIT observations of an Earth-directed coronal mass ejection on May 12, 1997. *Geophys. Res. Lett.* **25**, 2465–2468.
31. Wang H, Shen C, Lin J. 2009 Numerical Experiments of Wave-like Phenomena Caused by the Disruption of an Unstable Magnetic Configuration. *Astrophys. J.* **700**, 1716–1731.
32. Wills-Davey MJ, DeForest CE, Stenflo JO. 2007 Are "EIT Waves" Fast-Mode MHD Waves?. *Astrophys. J.* **664**, 556–562.
33. Vršnak B, Cliver EW. 2008 Origin of Coronal Shock Waves. Invited Review. *Solar Phys.* **253**, 215–235.
34. Chen PF, Wu ST, Shibata K, Fang C. 2002 Evidence of EIT and Moreton Waves in Numerical Simulations. *Astrophys. J. Lett.* **572**, L99–L102.
35. Delannée C, Török T, Aulanier G, Hochedez JF. 2008 A New Model for Propagating Parts of EIT Waves: A Current Shell in a CME. *Solar Phys.* **247**, 123–150.
36. Attrill GDR, Harra LK, van Driel-Gesztelyi L, Démoulin P. 2007 Coronal "Wave": Magnetic Footprint of a Coronal Mass Ejection?. *Astrophys. J. Lett.* **656**, L101–L104.
37. Biesecker DA, Myers DC, Thompson BJ, Hammer DM, Vourlidas A. 2002 Solar Phenomena Associated with "EIT Waves". *Astrophys. J.* **569**, 1009–1015.
38. Chen PF. 2009 The Relation Between EIT Waves and Coronal Mass Ejections. *Astrophys. J. Lett.* **698**, L112–L115.
39. Xu L, Liu S, Yan Y, Zhang W. 2020 EUV Wave Detection and Characterization Using Deep Learning. *Solar Phys.* **295**, 44.
40. He K, Gkioxari G, Dollár P, Girshick R. 2017 Mask r-cnn. In *Proceedings of the IEEE international conference on computer vision* pp. 2961–2969.
41. Simonyan K, Zisserman A. 2015 Very Deep Convolutional Networks for Large-Scale Image Recognition. In *International Conference on Learning Representations*.
42. Chollet F. 2019 github repository. https://github.com/fchollet/deep-learning-models/releases. Accessed May 15, 2019.
43. Center CD. 2019 LASCO CME Catalog. https://cdaw.gsfc.nasa.gov/CME_list/. Accessed June 28, 2019.
44. Gopalswamy N, Yashiro S, Michalek G, Stenborg G, Vourlidas A, Freeland S, Howard R. 2009 The SOHO/LASCO CME Catalog. *Earth, Moon, and Planets* **104**, 295–313.

Chapter 5
Deep Learning in Solar Image Generation Tasks

Abstract It has been witnessed that deep learning has been applied to classification in previous chapters. In fact, deep learning also demonstrated great ability of image generation which is more challenging than classification. In this chapter, several applications of deep learning in solar image enhancement, reconstruction and processing are presented, including image deconvolution of solar radioheliograph, desaturation of solar imaging, generating magnetogram, image super-resolution. These tasks are all concerned with image generation, by employing generative neural networks. As a representative of generative networks, GAN was widely exploited in image generation tasks. It can generate high fidelity and photo-realistic content mainly owning to an adversarial loss.

Keywords Image deconvolution · Aperture synthesis · Image generation · Magnetogram

5.1 Image Deconvolution of Solar Radioheliograph

For a single-dish antenna, its spatial resolution is determined by the dish diameter. It is difficult to construct a large single-dish antenna, so its spatial resolution can not be large. With aperture synthesis (AS) technique, a group of small antennas are assembled to form a large telescope with the spatial resolution given by the distance of two farthest antennas instead of the diameter of a single dish. Different from direct imaging system, an AS telescope captures the Fourier coefficients of a spatial image, and then implement inverse Fourier transform to reconstruct the spatial image. In practice, limited by the number of antennas, Fourier coefficients are extremely sparse, resulting in blur image. To remove/reduce blur, "CLEAN" deconvolution was developed. However, it was initially designed for point source. For extended source, like the sun, its efficiency is compromised. Inspired by the success of deep learning in image generation, a deep network based on GAN was proposed for solar image deconvolution. It was trained on the pair of original image and degraded one. After multiple rounds of "fighting" between a discriminator and a generator, a powerful generator can be acquired finally for image deconvolution.

5.1.1 Aperture Synthesis Principle

The imaging system of AS is demonstrated in Fig. 5.1, where $I(x, y)$ and $V(u, v)$ are brightness and visibility functions, corresponding original image in spatial and frequency domains, respectively. In practice, using AS, only a sparsely sampled $V(u, v)$, denoted by $V^D(u, v)$, can be obtained via

$$V^D(u, v) = V(u, v) \times S(u, v), \tag{5.1}$$

where $S(u, v)$ is a sampling function in frequency domain, corresponding to a dirty beam (or point spread function (PSF)) denoted by $B^D(l, m)$ in spatial domain. Applying inverse Fourier transform to (5.1), we can get

$$I^D(l, m) = I(l, m) \otimes B^D(l, m), \tag{5.2}$$

where the symbol "$\otimes$" denotes convolution operator.

From (5.2), an AS telescope only provides dirty image $I^D(x, y)$, while original image $I(x, y)$ is polluted by dirty beam $B(x, y)$. To restore $I(x, y)$ from (5.2),

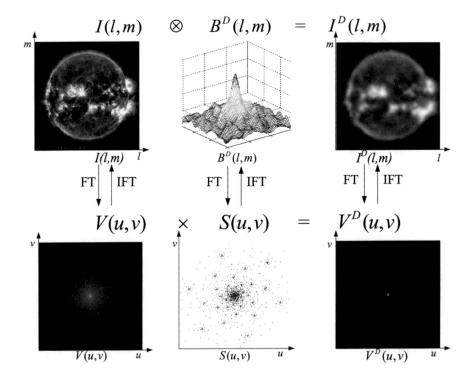

Fig. 5.1 Imaging principle of Aperture Synthesis

"CLEAN" deconvolution was raised, it eliminates the effect of dirty beam $B(x, y)$ from the left side of (5.2) iteratively. The first CLEAN algorithm was proposed by Högbom, et. al., namely Högbom CLEAN [1]. Its efficiency on point source has been well acknowledged, e.g., deconvolution for stellar objects. However, it was unsatisfied for extended source, like the Sun, so a bunch of variants of Högbom CLEAN were proposed, including multi-resolution CLEAN (MRC), multi-scale CLEAN and wavelet CLEAN [2, 3].

5.1.2 Proposed Model for Image Deconvolution

The principle of the proposed model for image deconvolution is demonstrated in Fig. 5.2. The baseline of the proposed model is a cGAN, specifically the pix2pix network described in [4]. Besides cGAN loss and L1 loss of spatial domain($\mathcal{L}_{L1}^{I}(G) = \mathbb{E}_{x,y,z}[\|y - G(x, z)\|_1]$), a new loss, namely perceptual loss [5], is also introduced additionally as,

$$\mathcal{L}_{L1}^{P}(G) = \mathbb{E}_{x,y,z}[\|\Phi(y) - \Phi(G(x, z))\|_1] \tag{5.3}$$

where $\Phi(\cdot)$ represents feature extraction given by a pre-trained VGG-16 model [6]. Here, we extract the first four layers of VGG-16 to give $\Phi(y)$ and $\Phi(G(x, z))$. Thus, the final objective is

$$G^* = \arg\min_{G}\max_{D} \mathcal{L}_{cGAN}(G, D) + \lambda_1\mathcal{L}_{L1}^{I}(G) + \lambda_2\mathcal{L}_{L1}^{P}(G) \tag{5.4}$$

In our model as shown in Fig. 5.3, the generator is a classical UNet, consisting of multiple layers of convolution and transposed convolution. The encoder gets compressed representation of the input, while the decoder decompresses this representation to reconstruct the input. The skip connection combines both high level semantic information and low level features of an image, benefiting

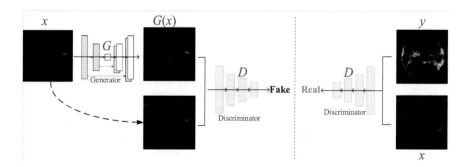

Fig. 5.2 Framework of image deconvolution network

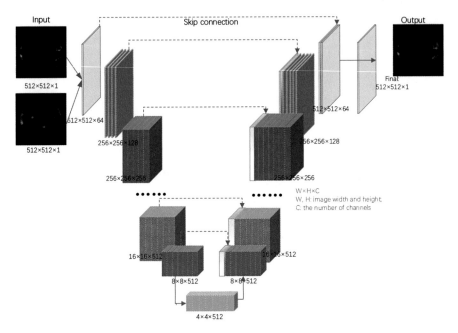

Fig. 5.3 Network architecture for AS image deconvolution

image generation tasks. The discriminator is a general convolution neural network consisting of 5 convolution layers.

Image generation/reconstruction, like image deblurring, denoising and super-resolution, has been well investigated in the literature [7–10]. Image deconvolution is a typical image generation problem. Usually, in radio astronomy, it was handled by CLEAN algorithm [1–3]. Two conditions should be held for the success of the CLEAN algorithm on image deconvolution. One is that the signal should be point source, the other is that dirty beam should be exactly known, which means the dirty beam of actual system and the ideal one are exactly the same. However, in practice, these two conditions do not hold so that the efficiency of the CLEAN algorithm is compromised.

5.1.3   Evaluation

To evaluate the proposed model, a database consisting original/clear and dirty image pairs is firstly established. We collected 41,096 images of 193 Å from Atmospheric Imaging Assembly (AIA) onboard SDO as ground-truth/clear images. Then, we apply MUSER-I dirty beam (as shown in Fig. 5.1b) to these clear images, resulting in corresponding dirty images. For training, validation and testing, the database is split into 3 parts: 8000 image pairs for validation, 8000 image pairs for testing. The full implementation (based on Pytorch) and the trained network can be accessed via https://github.com/filterbank/solarGAN.

After about 5000 loops, the learnt model can be stable, generating high quality images as shown in Fig. 5.4, where the left column gives dirty images, the middle column shows output images after GAN deconvolution, the right column shows original images. The learnt model can restore image details/structures well, as

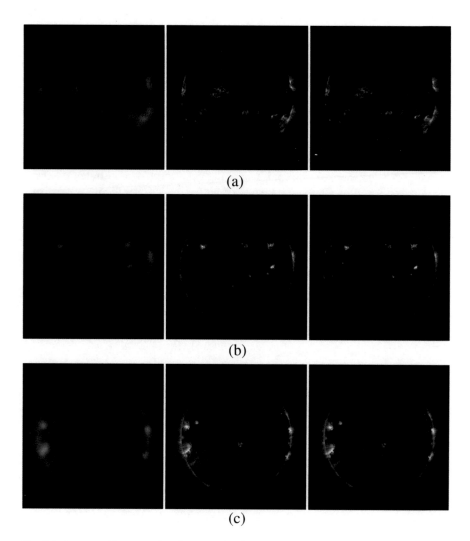

(a)

(b)

(c)

Fig. 5.4 Image quality comparison between dirty images (left), deconvolved images (middle) and original one (right) (SDO/AIA, 193 Å, dirty images are derived from MUSER-I sampling). (**a**) 2014-09-17, 09:00 (from left to right: dirty, deconvolved and original images). (**b**) 2017-02-01, 03:48 (from left to right: dirty, deconvolved and original images). (**c**) 2014-09-17, 09:12 (from left to right: dirty, deconvolved and original images)

shown in Fig. 5.4b. Compared with dirty image in Fig.5.4a, the reconstructed one contains more details of an image. We also verify the effectiveness of spatial loss and perceptual loss as claimed in (5.3) for our task. The peak signal to noise ratio (PSNR) and structural similarity index measurement (SSIM) [11] on the whole testing dataset are compared in Table 5.1. It can be observed that the best result is coming from the combination of cGAN loss, spatial-domain L1 loss and perceptual L1 loss.

For objective measurement of image quality, PSNR and SSIM are employed for evaluating the proposed model. PSNR measures the absolute difference of pixel-to-pixel of two images. SSIM may ignore the pixel-to-pixel difference, while pays more attention to the similarity of image structure. The PSNR and SSIM statistics are listed in Table 5.2. From Table 5.2, the remarkable improvements of image quality can be achieved by the proposed model, where the average PSNR improvement of 4.97 dB and average SSIM improvement of 5.2% are achieved by the proposed GAN-based model.

For comparison between the proposed model and traditional Högbom CLEAN, the dirty image in Fig. 5.4a is processed by Högbom CLEAN. The results of Högbom CLEAN are shown in Fig. 5.5, where Fig. 5.5a and b demonstrate the images of bright points after Högbom CLEAN of 400 and 4000 iterations, respectively. Figure 5.5c is the residual image corresponding to Fig. 5.5b. Figure 5.5d gives the final deconvoluted image which combines the residual image (Fig. 5.5c) and the

Table 5.1 Performance verification of the proposed network with different loss function (bold numbers indicate the best performance)

Loss function	PSNR	SSIM
$\mathcal{L}_{cGAN(G,D)} + \mathcal{L}_{L1}^{I}$	38.0575	0.9561
$\mathcal{L}_{cGAN(G,D)} + \mathcal{L}_{L1}^{I} + \mathcal{L}_{L1}^{P}$	**38.4442**	**0.9609**
$\mathcal{L}_{cGAN(G,D)} + \mathcal{L}_{L2}^{I}$	35.2378	0.9316
$\mathcal{L}_{cGAN(G,D)} + \mathcal{L}_{L2}^{I} + \mathcal{L}_{L2}^{P}$	37.3543	0.94727

Table 5.2 Performance evaluation of the proposed AS image deconvolution algorithm

Test images	PSNR(dB)		SSIM(block:8 × 8)		SSIM(block:16 × 16)	
	Deconvolved	Dirty	Deconvolved	Dirty	Deconvolved	Dirty
2012-08-31 19:48:06UT	43.4474	38.1938	0.9512	0.8965	0.9493	0.8932
2014-09-17 08:48:06UT	43.7925	38.3835	0.9527	0.9004	0.9509	0.8971
2014-09-17 09:00:06UT	43.7546	38.2539	0.9520	0.8975	0.9502	0.8942
2014-09-17 09:12:06UT	42.9609	37.6210	0.9508	0.8887	0.9491	0.8847
2015-05-28 12:48:06UT	43.3054	38.5841	0.9530	0.9028	0.9512	0.8988
2016-05-18 02:00:05UT	43.2166	38.5393	0.9531	0.9020	0.9512	0.8981
2016-05-18 02:12:05UT	43.2083	38.5592	0.9504	0.9014	0.9484	0.8979
2017-02-01 03:48:04UT	43.1904	38.4375	0.9500	0.8987	0.9481	0.8951
2017-02-01 04:00:04UT	44.0610	39.4934	0.9574	0.9174	0.9558	0.9140
2017-09-03 00:48:04UT	44.8305	40.0187	0.9599	0.9252	0.9584	0.9223
Average	**43.5768**	**38.6084**	**0.9531**	**0.9031**	**0.9513**	**0.8995**
Improvement	**4.9683**		**0.0500**		**0.0517**	

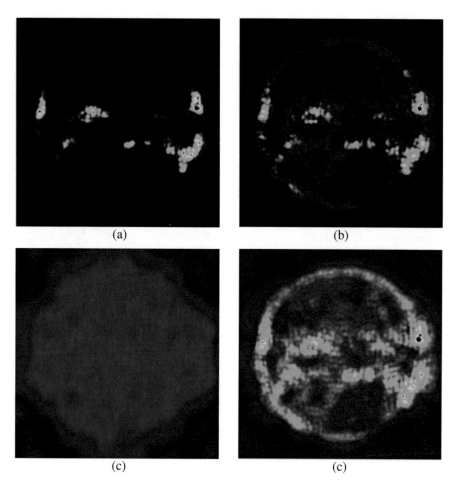

Fig. 5.5 The reconstructed image by using Högbom CLEAN ((**a**) and (**b**) only shows bright points without quiet solar background; here only grayscale images are processed since Högbom CLEAN is implemented on grayscale image). (**a**) Bright points of 400 iterations. (**b**) Bright points of 4000 iterations. (**c**) Residual image after 4000 iterations. (**d**) Final deconvoluted image by Högbom CLEAN

image of bight points (Fig. 5.5b). From Fig. 5.5, it can be concluded that Högbom CLEAN can successfully restore bright points in an image, however fail to restore image details. This conclusion also confirms that Högbom CLEAN is designed for point source instead of extended source. Comparing Figs. 5.5 and 5.4, the proposed model is dramatically superior to Högbom CLEAN on restoring image details/fine structures.

5.2 Recovery of Over-Exposed Solar Image

Over-exposure may happen for imaging of solar observation in case extremely violet solar bursts occur, which means that signal intensity goes beyond the dynamic range of an imaging system, resulting in information loss. Although over-exposure can be alleviated a little by reducing exposure time in case of flares, it cannot be solved completely. For example, during solar flare, SDO/AIA [12–15] often records over-exposed images/videos, resulting in loss of fine structures of solar flare.

For over-exposure region (OER) recovery (or desaturating), an iterative algorithm was introduced in [16] by employing the PRiL approximation [17] and EM algorithm. It recovered saturated region by referring to normal regions within an image. Recently, thanks to deep learning, lots of traditional image processing/reconstruction problems got breakthroughs, including image inpainting. Over-exposure recovery is similar to image inpainting task. The difference between them is twofold. First, some of the learning based methods, such as [18–21], and [22], trained their networks with fixed shapes and locations of missing regions, while over-exposure commonly has irregular shape and random location. Second, despite some efforts for irregular missing region of image inpainting, such as [23–25] and [26], they generate visually coherent completion or producing semantically plausible results, while OER recovery aims to restore missing information with high fidelity, besides visually plausible.

5.2.1 Proposed Mask-pix2pix Network

To address the aforementioned challenges, a learning-based model, namely mask-pix2pix network [27], was proposed to recover/complete over-exposure region. The proposed model is established over pix2pix [28], so it has the form of a GAN, where the generator and discriminator are an U-net [29] and a PatchGAN [30], respectively. Different from traditional pix2pix, the proposed model utilizes the Convolution-SwitchNorm-LReLU/ReLU [31] modules (LReLU for encoder and ReLU for decoder) rather than the Convolution-BatchNorm-ReLU [32]. The former (i.e. switchable normalization) can switch between BatchNorm [32], LayerNorm [33] and InstanceNorm [34] by learning their weights end-to-end. This improvement boosts the robustness of proposed model. In addition, loss function of the proposed model contains an adversarial cGAN loss, a masked L1 loss and an edge mask loss/smoothness. The adversarial cGAN loss can capture the full entropy of concerned conditional distributions, and thereby produce highly realistic textures. The masked L1 loss calculates the L1 loss only over OERs, enforcing correctness at low frequencies which guarantees restoration of high fidelity for OERs. The edge mask loss is used for smoothing edges of OERs and suppressing edge artifacts in the final restored image.

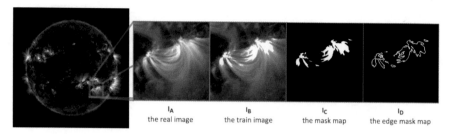

| I_A | I_B | I_C | I_D |
| the real image | the train image | the mask map | the edge mask map |

Fig. 5.6 The over-exposure region (OER) database built on the LSDO

In [27], a new database of over-exposure, collecting 13700 images from Large-scale Solar Dynamics Observatory image database (LSDO) [35], was established for training model. As shown in Fig. 5.6, ground truth image I_A, over-exposure image I_B, binary mask map I_C and edge mask map I_D are concerned with our task, recovering the region labeled by I_C in I_B, while keeping outside of I_C unchanged. Given OER Ω_M and non-OER $\Omega_{\overline{M}}$, we can get $\Omega_M = I_B \odot I_C$ and $\Omega_{\overline{M}} = I_B \odot (1 - I_C)$, where $\odot$ is the element-wise product operator. Inspired by image inpainting, the neural network, GAN, can be employed to retrieve missing region of I_B (i.e., Ω_M). In a GAN, a generator G is trained on the pairs of ground truth and degraded ones. Then, the generator is applied to a degraded image to output a repaired one, i.e., $I_G = G(I_B)$.

For our task, I_G is expected to contain realistic textures as far as possible, i.e. visually coherent and semantically plausible relative to I_A. In addition, $I_G \odot I_C$ has high fidelity relative to corresponding regions in ground truth I_A. Moreover, boundary between OERs and normal regions should transit smoothly to suppress artificial edges. To achieve these purposes, a mask L1 loss term and a edge mask loss/smoothness term are additionally introduced in loss function.

In the proposed mask-pix2pix model, the generator employs an U-Net architecture as demonstrated in Fig. 5.7. The U-Net is named after its shape which looks like a "U". It consists of an encoder of 8 layers, and a decoder of 8 layers. The parameters of each layer is explained in Table 5.3 in detail. In addition, skip connections are added between encoder and decoder at the same layer as shown in Fig.5.7 with dotted lines. Each skip connection simply concatenates feature map of encoder with that of decoder at the same layer (e.g., D_l and U_{n_l} at the l-th layer in Fig. 5.7). This cross-layer connection can reduce sematic gap between the same layers of encoder and decoder since they are far away in an U-Net structure. The discriminator is a PatchGAN network, the structure of which is explained in Table 5.4. In a GAN framework, the discriminator is to judge "fake" instances from "real" ones. In our work, the output of the discriminator is a 16×16 image, each pixel value of which ranges from 0 to 1, for measuring how real is the output. In addition, we adopt Convolution-SwitchNorm-LReLU/ReLU [31] instead of Convolution-BatchNorm-ReLU [32] of conventional pix2pix [28] in the proposed model. The former was proved to be more robust.

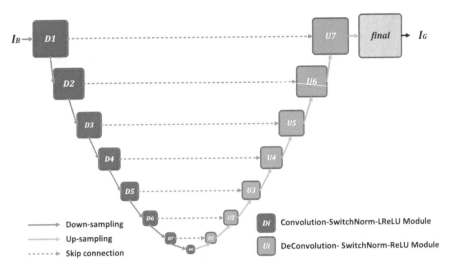

Fig. 5.7 Architecture of the generator

The OERs recovery is an image impainting task, but more concerns fidelity of reconstructed signal, natural transition between OERs and non-OERs. Therefore, a new hybrid loss function is designed for OER recovery task as follows.

Given ground truth image I_A, degraded image I_B, initial binary mask I_C (1 for OERs) and I_C's edge map I_D, as demonstrated in Fig. 5.6, the generator outputs $I_G = G(I_B)$. For training model, a hybrid loss function is defined, consisting of three components: (1) adversarial cGAN loss for high fidelity of reconstructed image I_G relative to ground-truth I_A, by referring to (2.7), which is given by

$$\mathcal{L}_{cGAN}(G, D) = \mathbb{E}_{I_B, I_A}[\log D(I_B, I_A)] + \mathbb{E}_{I_B, z}[\log(1 - D(I_B, G(I_B, z)))], \tag{5.5}$$

where G and D denote the generator and discriminator, respectively, z represents Gaussian noise, objective of G is to minimize $D(I_B, G(I_B, z))$ so that $G(I_B, z)$ looks more like a real image, while D maximizes $D(I_B, I_A)$ to distinguish "fake" and "real" as good as possible. Figure 5.8 illustrates the concept of adversarial cGAN training, consisting of a generator and a discriminator. In Fig. 5.8, the discriminator has two missions, checking whether I_A and I_B are an image pairs ("real" and "fake"), and whether $G(I_B)$ is "real" (or "fake"), while G produces "fake" images as real as possible to deceive D.

(2) L1 loss of masked region $I_B \odot I_C$ relative to $I_A \odot I_C$, for high accuracy of reconstructed OER, which is defined as:

$$\mathcal{L}_1^m(G) = \mathbb{E}[\|(I_A - G(I_B)) \odot I_C\|_1]. \tag{5.6}$$

Table 5.3 Architecture of the generator

Layers	Architecture of Generator	Output size
input	I_A, I_B, I_C, I_D	($256\times256\times1$)
D1	Conv.($4\times4\times64$), **LReLU**	($128\times128\times64$)
D2	Conv.($4\times4\times128$), **SwitchNorm, LReLU**	($64\times64\times128$)
D3	Conv.($4\times4\times256$), **SwitchNorm, LReLU**	($32\times32\times256$)
D4	Conv.($4\times4\times512$), **SwitchNorm, LReLU**, Dropout	($16\times16\times512$)
D5	Conv.($4\times4\times512$), **SwitchNorm, LReLU**, Dropout	($8\times8\times512$)
D6	Conv.($4\times4\times512$), **SwitchNorm, LReLU**, Dropout	($4\times4\times512$)
D7	Conv.($4\times4\times512$), **SwitchNorm, LReLU**, Dropout	($2\times2\times512$)
D8	Conv.($4\times4\times512$), **LReLU**, Dropout	($1\times1\times512$)
U1	Concatenate(D8, D7), DeConv.($4\times4\times512$), **SwitchNorm**, ReLU, Dropout	($2\times2\times512$)
U2	Concatenate(U1, D6), DeConv.($4\times4\times512$), **SwitchNorm**, ReLU, Dropout	($4\times4\times512$)
U3	Concatenate(U2, D5), DeConv.($4\times4\times512$), **SwitchNorm**, ReLU, Dropout	($8\times8\times512$)
U4	Concatenate(U3, D4), DeConv.($4\times4\times512$), **SwitchNorm**, ReLU, Dropout	($16\times16\times512$)
U5	Concatenate(U4, D3), DeConv.($4\times4\times256$), **SwitchNorm**, ReLU	($32\times32\times256$)
U6	Concatenate(U5, D2), DeConv.($4\times4\times128$), **SwitchNorm**, ReLU	($64\times64\times128$)
U7	Concatenate(U6, D1), DeConv.($4\times4\times64$), **SwitchNorm**, ReLU	($128\times128\times64$)
Final	Upsample($4\times4\times1$), ZeroPad, Conv.($4\times4\times1$), *tanh*	($256\times256\times1$)
Output	I_G	($256\times256\times1$)

Table 5.4 Architecture of the discriminator

Layers	Architecture of discriminator	Output size
input	$[I_A\ I_B]$ or $[I_G^*\ I_B]$	($256\times256\times2$)
C1	Conv.($4\times4\times64$), **LReLU**	($128\times128\times64$)
C2	Conv.($4\times4\times128$), **SwitchNorm, LReLU**	($64\times64\times128$)
C3	Conv.($4\times4\times256$), **SwitchNorm, LReLU**	($32\times32\times256$)
C4	Conv.($4\times4\times512$), **SwitchNorm, LReLU**	($16\times16\times512$)
C5	ZeroPad2d, Conv.($4\times4\times1$)	($16\times16\times1$)
Output	Real or Fake matrix 16×16	

(3) L1 loss of edge mask, which could make the edges of OERs smooth to prevent artificial edges connecting OERs and non-OERs. It is defined as

$$\mathcal{L}_1^e(G) = \mathbb{E}[\|(I_A - G(I_B)) \odot I_D\|_1]. \tag{5.7}$$

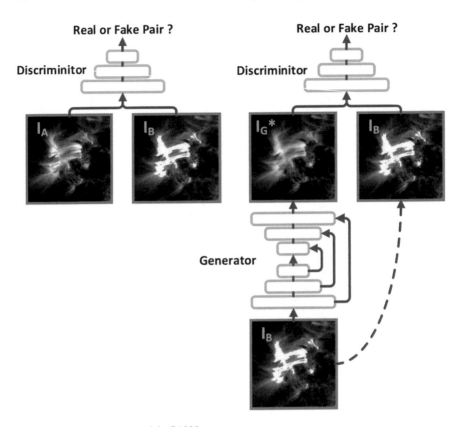

Fig. 5.8 Training on adversarial cGAN loss

In conclusion, the final optimization objective is given by

$$G^* = \arg \min_{G} \max_{D} \mathcal{L}_{cGAN}(G, D) + \lambda_1 \mathcal{L}_1^m(G) + \lambda_2 \mathcal{L}_1^e(G), \qquad (5.8)$$

where λ_1 and λ_2 are the weights for combining the above three loss components. They are set to 0.1 in our experience.

5.2.2   Evaluations

To verify the proposed mask-pix2pix model, it is implemented by PyTorch on our raised database, and compared with other state-of-the-art image inpainting algorithms. In addition, ablation experiments are performed to evaluate each component of the loss function. We train our model using the Adam optimizer, where $\beta_1 = 0.5$, $\beta_2 = 0.999$, and $\varepsilon = 10^{-8}$. The image resolution is 256×256, and batch size is 16. The initial learning rate is initialized to 0.0002, and then reduced to half every 100 epochs.

5.2.2.1 Database

For model training, a database for over-exposure is built over LSDO [35] which consists of event records, corresponding images, and some image parameters. Event records encompass the list of solar events with generated and extracted attributes. There are four kinds of events, AR, filament, solar flare and sigmoid. For an AR, bounding box attributes give top-left and bottom-right coordinates of a polygon enclosing the AR, so corresponding AR can be cropped out from full-disk SDO/AIA images. The cropped ARs are scaled to the same resolution (512×512), denoted by I_A. Then, imposing a threshold on ARs to produce images with over-exposure, denoted by I_B. From I_B, we can extract a mask map, denoted by I_C, which is further dilated by a circular kernel (radius of 3). Final, edge mask map I_D accounting for edge mask loss/smoothness is extracted from I_C. Thereby, each sample in database contains four components: real image I_A, over-exposed (fake) image I_B, mask map I_C and edge mask map I_D.

5.2.2.2 Comparisons with State-of-the-Art Approaches

We compare the proposed mask-pix2pix with a patch-based Planar Structure Guidance (PSG) [36] and pix2pix [28]. Figure 5.9 shows visual comparisons on 3 samples. It can be found that the proposed mask-pix2pix can generate images with significant visual quality improvement over two benchmarks, PSG does not recover OERs sufficiently, pix2pix [28] may underestimate OERs, still leaving over-exposure OERs after reconstruction. We quantitatively measure their performances using peak signal-to-noise ratio (PSNR) and structural similarity index (SSIM) [11]. The PSNR and SSIM are shown below each reconstructed image in Fig. 5.9. Average PSNR and SSIM are also computed on the whole database (over 1600 samples), listed in Table 5.5. It can be observed that the proposed model outperforms other two benchmarks significantly, achieving PSNR gain up to 5 dB relative to the pix2pix [28] and 15 dB relative to the PSG [36].

5.2.2.3 Ablation Studies

To evaluate contribution of each component of loss function, we perform the following three ablation experiments:

(1) adversarial loss and conventional L1 loss on whole image, i.e. $\mathcal{L}_{cGAN}(G, D) + \lambda_1 \mathcal{L}_1(G)$;
(2) adversarial loss and masked L1 loss, i.e. $\mathcal{L}_{cGAN}(G, D) + \lambda_1 \mathcal{L}_1^m(G)$;
(3) adversarial loss, masked L1 loss and edge mask loss/smoothness, i.e. $\mathcal{L}_{cGAN}(G, D) + \lambda_1 \mathcal{L}_1^m(G) + \lambda_2 \mathcal{L}_1^e(G)$.

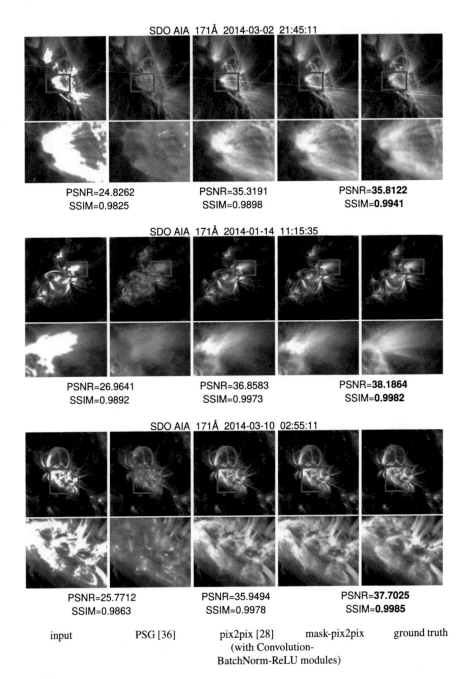

Fig. 5.9 Comparisons between the proposed one and two state-of-the-art methods

Table 5.5 Average PSNR and SSIM comparisons between our model and other two benchmarks (bold numbers indicate the best performance)

Methods	Loss functions	Modules	Average PSNR	Average SSIM
PSG [36]			24.6219	0.9810
pix2pix [28]	$\mathcal{L}_{cGAN} + \lambda_1 \mathcal{L}_1$ [a]	Convolution-BatchNorm-ReLU	34.7763	0.9891
pix2pix	$\mathcal{L}_{cGAN} + \lambda_1 \mathcal{L}_1$ [a]	Convolution-SwitchNorm-LReLU/ReLU	35.9939	0.9918
mask-pix2pix	$\mathcal{L}_{cGAN} + \lambda_1 \mathcal{L}_1^m$ [a]	Convolution-SwitchNorm-LReLU/ReLU	37.4334	0.9937
mask-pix2pix	$\mathcal{L}_{cGAN} + \lambda_1 \mathcal{L}_1^m + \lambda_2 \mathcal{L}_1^e$ [b]	Convolution-SwitchNorm-LReLU/ReLU	**39.6931**	**0.9985**

[a] $\lambda_1 = 0.1$.
[b] $\lambda_1 = \lambda_2 = 0.1$.

Figure 5.10 illustrates the experimental results of the three experiments. It can be found that the first ablation experiment fails to estimate accurate intensity and content of missed region, despite of smooth boundary between OERs and non-OERs. Benefiting from the proposed masked loss, the second ablation experiment could well address the problem of the first one, however, it yields artificial edges as illustrated in the first and third images in Fig. 5.10. Importing edge mask loss $\mathcal{L}_1^e$ beyond the second one, the third experiments, namely the proposed one, could well address the above two problems.

5.3   Generating Magnetogram from EUV Image

Magnetic filed manipulates all of solar activities occurring in solar atmosphere. Solar bursts, such as flares and CMEs, happening in high solar atmosphere, are still highly associated with magnetic field of photosphere in low solar atmosphere. Exploring the relation between magnetic filed of photosphere and EUV observation of corona is great importance for us to understand the mechanism of solar bursts.

SDO/AIA [37] provides UV and EUV observatories of solar chromosphere and corona. It can present rich solar activities of chromosphere and corona. SDO/HMI can measure photospheric magnetic fields. Galvez et al. [38] pointed that the mapping between HMI and AIA from physical mechanism. Kim et al. [39] thought of this mapping as an image-to-image translation, generating farside magnetograms from STEREO [40] EUV 304 Å observatories by using pix2pix network [41]. The model of [39] can provide the entire evolution of an active region from the farside to the frontside of the Sun.

We reproduced Kim's algorithm, and compared SDO/HMI magnetograms and generated magnetograms in Fig. 5.11. It can be observed that the model successfully generates frontside magnetograms from SDO/AIA 304 Å images. The generated

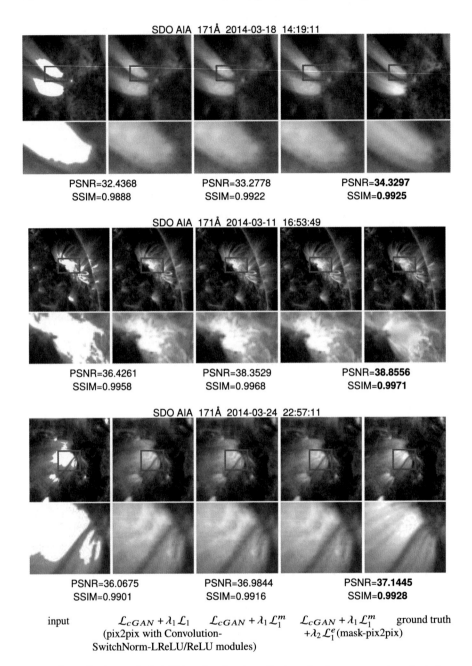

Fig. 5.10 Evaluation on the hybrid loss function (contribution of each component is evaluated)

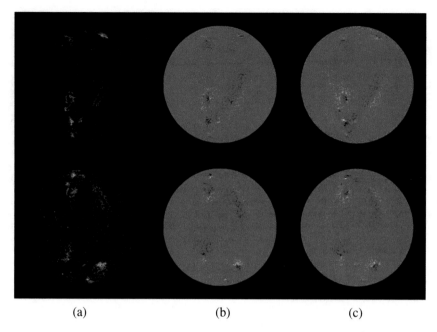

(a)   (b)   (c)

Fig. 5.11 Generated Magnetograms from SDO/AIA. (**a**) SDO/AIA 304 Å. (**b**) Generated magnetograms. (**c**) SDO/HMI magnetograms

images are similar to those of SDO/HMI in morphology, with active regions, polarities being well aligned. More importantly, we can generate magnetograms of the farside of the Sun when applied the well trained model on SDO to STEREO as shown in Fig. 5.12. As we know, STEREO consists of two nearly identical satellites, one ahead of the Earth in its orbit and the other trailing behind, to provide the first-ever stereoscopic measurements to study the Sun and the nature of its coronal mass ejections, or CMEs. Unfortunately, it did not carry a magnetograph. Using Kim's model, it is possible to generate magnetograms from SECCHI EUVI (Extreme UltraViolet Imager) onboard STEREO. Thus, we can monitor the continuous evolution of magnetic fields from the farside to the frontside of the Sun when the farside EUV observation is available, cooperating with SDO/HMI.

However, it suffered from magnetic pole reversal problem, inconsistent motion and fluctuating visual perception for the sequence of magnetograms. The reason lies in that Kim's model [39] was a static model established on a single input and single output system, so it cannot well describe the gradual transition and stable geometric construction of an input sequence.

For processing a time series, we add a convolutional Gated Recurrent Units(GRU) [42] to a general pix2pix GAN [41] to construct a convGRU-pix2pix network which is a dynamic model. A dynamic model could explore sequential information from input image sequence beside mining information from each individual image, benefiting for generating consistent magnetogram sequence.

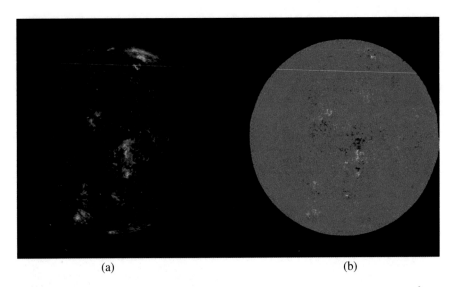

Fig. 5.12 Generated Magnetograms from EUV observation of STEREO. (**a**) STEREO 304 Å. (**b**) Generated magnetogram from (**a**)

In addition, a dynamic model could bridge the gap between source domain and target domain in case their variations are asynchronous. Compared to HMI, AIA observatories record solar activities at high solar atmosphere, where fast local brightening or mass ejection usually happen. The evolution of HMI and AIA is asynchronous. The former evolves slowly, even solar bursts happening. While the later changes fast and violent, especially in the case of solar bursts.

The baseline of the proposed model is a general pix2pix GAN, consisting of a generator and a discriminator. The generator is a typical U-Net, i.e., an encoder-decoder structure. As shown in Fig. 5.13, a basic U-net is equipped with a convGRU module between encoder and decoder, granting the ability of processing time series over a static model.

GRU was proposed by Cho et al. [43], being a RNN model, which allows each recurrent unit to capture sequence dependencies over various lengths of time adaptively. Compared to LSTM, GRU combines cell state and hidden state, resulting in a lightweight model. The GRU has a gating structure which controls the flow of information inside the unit and is formulated as follows:

$$r_t = \delta(W^r x_t + R^r h_{t-1} + b^r) \tag{5.9}$$

$$z_t = \delta(W^z x_t + R^z h_{t-1} + b^z) \tag{5.10}$$

$$\tilde{h}_t = \varphi(W^h x_t + r_t \odot (R^h h_{t-1}) + b^h) \tag{5.11}$$

$$h_t = z_t \odot h_{t-1} + (1 - z_t) \odot \tilde{h}_t \tag{5.12}$$

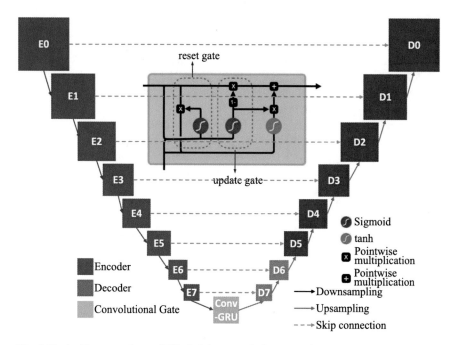

Fig. 5.13 Architecture of convGRU-pix2pix network for generating magnetogram from EUV image sequence

where $\tilde{h}_t$ and h_t are candidate hidden state and hidden state at time t respectively, r_t and z_t are reset gate and update gate at time t respectively, W^* and R^* ($* = r, z, h$) denote the weights of network, b^* ($* = r, z, h$) represents bias of network. The nonlinear functions $\delta(\cdot)$ and $\varphi(\cdot)$ are sigmoid and *tanh* functions, respectively.

Following the work of [39], SDO/AIA 304 Å and SDO/HMI magnetogram form image pairs for model training. Three AIA 304 Å sequential images and one HMI image are grouped together to provide a sample, where three AIA sequential images constitute a conditional input, and one HMI magnetogram is ground-truth to the proposed convGRU-pix2pix. The HMI magnetogram has the same time stamp with the third AIA 304 Å image. There are about 40 thousands samples in total, constituting a database.

We compares three models, pix2pix, multi-channel pix2pix and convGRU pix2pix, to verify the effectiveness of time series model for generating magnetograms. It can observed that convGRU pix2pix model has the most consistent evolution of magnetic field, especially around strong magnetic field from generated magnetogram sequence https://github.com/filterbank/solarMag.

5.4 Generating Magnetogram from Hα

Hα is a specific deep-red visible spectral line created by hydrogen with a wavelength of 6562.8 nm. Thus, Hα telescope can be placed on the ground. Since there is close correlation between magnetogram and Hα, we propose to learn mapping between Hα and HMI magnetogram by using cGAN.

Mapping between SDO/HMI and Hα images concerns an image-to-image translation task. Pix2pix model was raised for image-to-image translation. Its baseline is a cGAN, consisting of a generator and a discriminator. The objective function of a cGAN is described by (2.7) concerning an adversarial loss. Besides, L1 loss measuring fidelity of signal is employed in our model. The proposed model takes the same framework to pix2pix, while optimization objective is different from pix2pix. As mentioned above, static model, like pix2pix, cannot well fit for the task of generating magnetogram from SDO/AIA image. The reason lies in that SDO/AIA images and SDO/HMI magnetograms are asynchronous at the same time scale. Surprisedly, this static model performs well on generating magnetogram from Hα since both of them record the images of photosphere, with the same scale of evolution of solar activities.

The pix2pix network adopts an U-Net as the generator and a patch-wised fully convolutional network as the discriminator. The details of network configuration are listed in Table 5.6. The encoder has eight blocks, each of which is composed of Convolution, InstanceNorm and LeakyReLu. Among these blocks, the forth to eighth ones take dropout with probability of 0.5. The decoder has eight blocks, each

Table 5.6 Network configuration used for magnetogram generation from Hα (Ei and Di (i = 0, 1, ..., 7) denote the i-th layer of encoder and decoder, respectively)

Modules		Conv2d	ConvTranpose	InstanceNorm	LeakyReLU	ReLU	Dropout
Encoder	E0	✓		✓	✓		
	E1	✓		✓	✓		
	E2	✓		✓	✓		
	E3	✓		✓	✓		0.5
	E4	✓		✓	✓		0.5
	E5	✓		✓	✓		0.5
	E6	✓		✓	✓		0.5
	E7	✓		✓	✓		0.5
Decoder	D0		✓	✓		✓	
	D1		✓	✓		✓	
	D2		✓	✓		✓	
	D3		✓	✓		✓	0.5
	D4		✓	✓		✓	0.5
	D5		✓	✓		✓	0.5
	D6		✓	✓		✓	0.5
	D7		✓	✓		✓	0.5

of which consists of Transposeconvolution, InstanceNorm and ReLu. In addition, the ninth to fourteenth blocks employ dropout with the probability of 0.5.

For model training, a database consisting of magnetogram and Hα pairs is established. The Hα images and magnetograms come from the Global Oscillation Network Group (GONG [44]) and HMI [45]) onboard SDO [37]), respectively. GONG includes six stations (the Big Bear Solar Observatory, High Altitude Observatory, Learmonth Solar Observatory, Udaipur Solar ObservatoryInstituto de Astrofísica de Canarias and Cerro Tololo Interamerican Observatory) located around all over the world. GONG provides full-disk line-center Hα solar images (2048 × 2048) with time cadences about 1 min. The line-of-sight (LOS) magnetic field is provided by SDO/HMI, observing solar photospheric magnetic field with high temporary and spatial resolutions. We collect 500 Hα and HMI LOS image pairs from 2012 to 2013. The original Hα and HMI images are 2048 × 2048 and 4096 × 4096, respectively.

We get paired Hα and HMI according to their timestamps. HMI has spatial resolution of 0.5 arcsec per pixel, while GONG Hα has only half of spatial resolution of HMI. For adapting to GPU ability, both of them are downsampled to 512 × 512. Finally, a database includes 654 paired images is obtained. During training, training set, validation set and test set have 394, 130 and 130 image pairs, respectively.

Figure 5.14 demonstrates an example of generating magnetogram from Hα for quantitative evaluation, where the left image is a Hα image, the middle one is a SDO/HMI image and the right one is the generated magnetogram. It can be observation that the generated magnetogram is very close to the ground-truth. In addition, it is found that generated magnetogram sequence changes smoothly without abrupt fluctuation which occurs to generating magnetogram from SDO/AIA. The reason lies in that Hα is more close to magetogram of photosphere, while EUV image of SDO/AIA behaves more active and violent, especially in case solar flares.

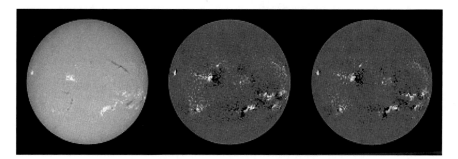

Fig. 5.14 Example of generating magnetogram from Hα

References

1. Högbom J. 1974 Aperture synthesis with a non-regular distribution of interferometer baselines. *Astronomy and Astrophysics Supplement Series* **15**, 417.
2. Wakker BP, Schwarz U. 1988 The Multi-Resolution CLEAN and its application to the short-spacing problem in interferometry. *Astronomy and Astrophysics* **200**, 312–322.
3. Cornwell TJ. 2008 Multiscale CLEAN deconvolution of radio synthesis images. *IEEE Journal of Selected Topics in Signal Processing* **2**, 793–801.
4. Isola P, Zhu JY, Zhou T, Efros AA. 2016 Image-to-Image Translation with Conditional Adversarial Networks. *arXiv e-prints*.
5. Johnson J, Alahi A, Fei-Fei L. 2016 Perceptual Losses for Real-Time Style Transfer and Super-Resolution. *Lecture Notes in Computer Science* pp. 694–C711.
6. Simonyan K, Zisserman A. 2015 Very Deep Convolutional Networks for Large-Scale Image Recognition. In *International Conference on Learning Representations*.
7. Kupyn O, Budzan V, Mykhailych M, Mishkin D, Matas J. 2017 DeblurGAN: Blind Motion Deblurring Using Conditional Adversarial Networks. *arXiv e-prints*.
8. KupynOrest. 2019 github repository. https://github.com/KupynOrest/DeblurGAN.git. Accessed Aug. 19, 2019.
9. Nah S, Kim TH, Lee KM. 2016 Deep Multi-scale Convolutional Neural Network for Dynamic Scene Deblurring. *arXiv e-prints*.
10. Yan Q, Wang W. 2017 DCGANs for image super-resolution, denoising and debluring. *Adv. Neural Inf. Process. Syst.* **8**, 487–495.
11. Wang Z, Bovik AC, Sheikh HR, Simoncelli EP et al. 2004 Image quality assessment: from error visibility to structural similarity. *IEEE transactions on image processing* **13**, 600–612.
12. Lemen JR, Akin DJ, Boerner PF, Chou C, Drake JF, Duncan DW, Edwards CG, Friedlaender FM, Heyman GF, Hurlburt NE et al. 2011 The atmospheric imaging assembly (AIA) on the solar dynamics observatory (SDO). In *The Solar Dynamics Observatory* pp. 17–40. Springer.
13. Pesnell WD, Thompson BJ, Chamberlin PC. 2011 The solar dynamics observatory (SDO). In *The Solar Dynamics Observatory* pp. 3–15. Springer.
14. Boerner P, Edwards C, Lemen J, Rausch A, Schrijver C, Shine R, Shing L, Stern R, Tarbell T, Wolfson CJ et al. 2011 Initial calibration of the atmospheric imaging assembly (AIA) on the solar dynamics observatory (SDO). In *The Solar Dynamics Observatory* pp. 41–66. Springer.
15. Scherrer PH, Schou J, Bush R, Kosovichev A, Bogart R, Hoeksema J, Liu Y, Duvall T, Zhao J, Schrijver C et al. 2012 The helioseismic and magnetic imager (HMI) investigation for the solar dynamics observatory (SDO). *Solar Physics* **275**, 207–227.
16. Guastavino S, Piana M, Massone AM, Schwartz R, Benvenuto F. 2019 Desaturating EUV observations of solar flaring storms. *arXiv preprint arXiv:1904.04211*.
17. Sabrina G, Federico B. 2018 A consistent and numerically efficient variable selection method for sparse Poisson regression with applications to learning and signal recovery. *Statistics and Computing* **29**, 1–16.
18. Yu F, Koltun V. 2015 Multi-scale context aggregation by dilated convolutions. *arXiv preprint arXiv:1511.07122*.
19. Pathak D, Krahenbuhl P, Donahue J, Darrell T, Efros AA. 2016 Context encoders: Feature learning by inpainting. In *Proceedings of the IEEE conference on computer vision and pattern recognition* pp. 2536–2544.
20. Yang C, Lu X, Lin Z, Shechtman E, Wang O, Li H. 2017 High-resolution image inpainting using multi-scale neural patch synthesis. In *Proceedings of the IEEE Conference on Computer Vision and Pattern Recognition* pp. 6721–6729.
21. Iizuka S, Simo-Serra E, Ishikawa H. 2017 Globally and locally consistent image completion. *ACM Transactions on Graphics (ToG)* **36**, 107.
22. Song Y, Yang C, Lin Z, Liu X, Huang Q, Li H, Jay Kuo CC. 2018 Contextual-based image inpainting: Infer, match, and translate. In *Proceedings of the European Conference on Computer Vision (ECCV)* pp. 3–19.

23. Liu G, Reda FA, Shih KJ, Wang TC, Tao A, Catanzaro B. 2018 Image inpainting for irregular holes using partial convolutions. In *Proceedings of the European Conference on Computer Vision (ECCV)* pp. 85–100.

24. Yan Z, Li X, Li M, Zuo W, Shan S. 2018 Shift-net: Image inpainting via deep feature rearrangement. In *Proceedings of the European Conference on Computer Vision (ECCV)* pp. 1–17.

25. Xiao Q, Li G, Chen Q. 2018 Deep Inception Generative Network for Cognitive Image Inpainting. *arXiv preprint arXiv:1812.01458.*

26. Nazeri K, Ng E, Joseph T, Qureshi F, Ebrahimi M. 2019 EdgeConnect: Generative Image Inpainting with Adversarial Edge Learning. *arXiv preprint arXiv:1901.00212.*

27. Zhao D, Xu L, Chen L, Yan Y, Duan LY. 2019 Mask-Pix2Pix Network for Overexposure Region Recovery of Solar Image. *Advances in Astronomy* **2019**, 5343254.

28. Isola P, Zhu JY, Zhou T, Efros AA. 2017 Image-to-image translation with conditional adversarial networks. In *Proceedings of the IEEE conference on computer vision and pattern recognition* pp. 1125–1134.

29. Ronneberger O, Fischer P, Brox T. 2015 U-net: Convolutional networks for biomedical image segmentation. In *International Conference on Medical image computing and computer-assisted intervention* pp. 234–241. Springer.

30. Mirza M, Osindero S. 2014 Conditional generative adversarial nets. *arXiv preprint arXiv:1411.1784.*

31. Luo P, Ren J, Peng Z. 2018 Differentiable learning-to-normalize via switchable normalization. *arXiv preprint arXiv:1806.10779.*

32. Ioffe S, Szegedy C. 2015 Batch normalization: Accelerating deep network training by reducing internal covariate shift. *arXiv preprint arXiv:1502.03167.*

33. Lei Ba J, Kiros JR, Hinton GE. 2016 Layer normalization. *arXiv preprint arXiv:1607.06450.*

34. Ulyanov D, Vedaldi A, Lempitsky V. 2016 Instance normalization: The missing ingredient for fast stylization. *arXiv preprint arXiv:1607.08022.*

35. Kucuk A, Banda JM, Angryk RA. 2017 A large-scale solar dynamics observatory image dataset for computer vision applications. *Scientific data* **4**, 170096.

36. Huang JB, Kang SB, Ahuja N, Kopf J. 2014 Image completion using planar structure guidance. *ACM Transactions on graphics (TOG)* **33**, 129.

37. SDO Obseratory. [EB/OL]. https://sdo.gsfc.nasa.gov/ Accessed Oct. 31, 2020.

38. Galvez R, Fouhey DF, Jin M, Szenicer A, Muñoz-Jaramillo A, Cheung MCM, Wright PJ, Bobra MG, Liu Y, Mason J, Thomas R. 2019 A Machine-learning Data Set Prepared from the NASA Solar Dynamics Observatory Mission. *The Astrophysical Journal Supplement Series* **242**, 7.

39. Kim T, Park E, Lee H, Moon YJ, Bae SH, Lim D, Jang S, Kim L, Cho IH, Choi M, Cho KS. 2019 Solar farside magnetograms from deep learning analysis of STEREO/EUVI data. *Nature Astronomy* **3**, 397–400.

40. Stereo Obseratory. [EB/OL]. https://stereo.gsfc.nasa.gov/ Accessed Oct. 31, 2020.

41. Isola P, Zhu JY, Zhou T, Efros AA. 2016 Image-to-Image Translation with Conditional Adversarial Networks.

42. Cho K, van Merriënboer B, Bahdanau D, Bengio Y. 2014a On the Properties of Neural Machine Translation: Encoder–Decoder Approaches. In *Proceedings of SSST-8, Eighth Workshop on Syntax, Semantics and Structure in Statistical Translation* pp. 103–111 Doha, Qatar. Association for Computational Linguistics.

43. Cho K, van Merriënboer B, Bahdanau D, Bengio Y. 2014b On the Properties of Neural Machine Translation: Encoder–Decoder Approaches. In *Proceedings of SSST-8, Eighth Workshop on Syntax, Semantics and Structure in Statistical Translation* pp. 103–111 Doha, Qatar. Association for Computational Linguistics.

44. Frank, Hill, George, Fischer, Jennifer, Grier, John, W., Leibacher, and H. 1994 The global oscillation network group site survey. *Solar Physics.*

45. 2011 Design and Ground Calibration of the Helioseismic and Magnetic Imager (HMI) Instrument on the Solar Dynamics Observatory (SDO). *Solar Physics.*

Chapter 6
Deep Learning in Solar Forecasting Tasks

Abstract Besides classification and generation, deep learning is also applicable to time series analysis. Unlike CNN which accepts singe image input, RNN is specifically designed for handling time series input, e.g., video sequence, natural language processing. As the best representative of RNN, LSTM has been widely exploited in various of time series analysis, achieving big success. In this chapter, it is applied to solar activity/event forecasting and solar radiation index prediction. As one of the most violent solar eruptions, solar flare is the main driving source of catastrophic space weather, so forecasting of solar flare is of great importance. The solar radio flux of 10.7 cm is a typical index for measuring global solar activity. It is a typical indicator of long-term space weather.

Keywords Solar flare forecasting · Recurrent neural network (RNN) · F10.7 cm radio flux · Solar cycle

6.1 Solar Flare Forecasting

Solar flares are caused by the release of energy stored in the magnetic field of solar active regions. Their triggering mechanism, however, remain unknown. For this reason, solar flare forecasting is highly dependent on statistics concerning relationship between solar flare and magnetic/topological parameters of active region. In our previous effort [1], deep learning method was employed to establish solar flare forecasting model over line-of-sight magnetograms of solar active regions. For validation of the proposed model, a database was constructed, containing magnetograms of active regions recorded by SOHO/MDI and SDO/HMI from April 1996 to October 2015. The experimental results indicated that the proposed model is comparable to the state-of-the-art models. In addition, the proposed model is robust, not sensitive to noise. Moreover, heat map of network demonstrated which characteristics of magnetic field contribute most to solar flare forecasting.

© The Author(s), under exclusive license to Springer Nature Singapore Pte Ltd. 2022
L. Xu et al., *Deep Learning in Solar Astronomy*, SpringerBriefs in Computer
Science, https://doi.org/10.1007/978-981-19-2746-1_6

6.1.1 History of Solar Flare Forecasting

Since 1930s, statistical model was widely used for modeling of solar flare forecasting [2–7]. With the increase of interesting in AI, machine learning was developed to discover the knowledge from massive data for solar flare forecasting, such as well-known SVM [8–10], neural network [11–14], ordinal logistic regression [15], Bayesian network [16], and ensemble learning [17]. Both statistical and conventional machine learning based models of solar flare forecasting rely on morphological and physical parameters of active regions. Compared to morphological parameters, physical parameters which measure complexity and non-potentiality of active regions are more reasonable, e.g., neutral line length [18], magnetic field gradient [19], highly stressed longitudinal magnetic field [20], distance between active regions and predicted active longitudes [21], Zernike moment of magnetograms [22]. Furthermore, The evolutions of physical parameters in active regions are studied in [23–27]. However, most of parameters extracted from active regions are strongly correlated with each other, resulting in comparable performance of the models using these parameters [6, 28, 29]. Extracting efficient parameters from action regions has become a bottleneck to solar flare forecasting for a long time.

Deep learning has been developed fast [30–32] in the last decade. The most exciting characteristic of deep learning is an end-to-end feature learning. Thus, it can automatically mine some specific forecasting patterns hidden in massive data, rather than extract hand-crafted physical/morphological parameters from active regions for solar flare forecasting, which has made a breakthrough in solar flare forecasting.

6.1.2 Deep Learning Model for Solar Flare Forecasting

Applying shallow models (e.g., SVM) to solar flare forecasting, physical parameters would be extracted from magnetograms at first. Then, a classifier/regressor is trained over these parameters. With deep learning, we now learn both features of magnetograms and a classifier/regressor simultaneously from an end-to-end network. In our previous effort [1], we proposed a solar flare forecasting model on the baseline of CNN, achieving state-of-the-art performance.

The framework of the proposed model is illustrated in Fig. 6.1, consisting of convolution layer, nonlinear layer, pooling layer and fully-connected layer. Each convolutional layer has 64 filters to ensure ability of extracting forecasting patterns from input data. The size of the filters is set to 11×11. Network parameters/weights are initialized by a zero-mean Gaussian function with 0.02 standard deviation. The gradient descent algorithm based optimization can not ensure the global optimization solution, Although different weights result in different solution of optimization objective, suboptimal solutions which can get the similar results for the different initializations of the random weights. To unify image size, images of arbitrary sizes are first revised into a fixed size via cropping or warping [33].

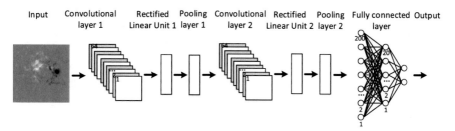

Fig. 6.1 Framework of the proposed solar flare forecasting model

Here, we process all images into size of 100×100. Then, convolution operation is performed in convolutional layer. After that, rectified linear unit is applied to achieve nonlinear transform. And then, max pooling is imposed on output of nonlinear layer to compress network volume, for avoiding over-fitting. On the top of convolution layers, three fully-connected layers are stacked. Finally, network outputs a probability, representing whether a flare event will happen.

After selecting network structure, weights of the network are learnt from massive data by using stochastic gradient descent (SGD) algorithm. The learning rate during SGD determines how fast the training process is converged. In the case of small learning rate, training process would converge slowly, otherwise training process may diverge. Here, learning rate is set to 0.01. The loss function is a cross-entropy between network outputs and labels of training samples, formulated by

$$L = -\sum_i y_i \log(x_i), \tag{6.1}$$

where y_i and x_i are labels and network outputs, respectively.

6.1.3   Model Validation

The proposed model was implemented over the Caffe platform [34], developed by the Berkeley Vision and Learning Center. First, we trained model over SOHO/MDI, and tested it over SDO/HMI. The forecasting accuracy measured by TPR and FPR is listed in Table 6.1. It can be observed that the proposed model can achieve TPR of 0.85 for M flare forecasting in 24 h. Second, we compared the proposed model with state-of-the-art methods. Since these models were trained and tested over different dataset, and with different event definition, forecasting cadence and forecasting period, we only taken 24 h M flare forecasting as an example in [1]. Murray et al. [35] carried out forecasting over full disk, where the dataset included 141 positive samples and 1489 negative samples from 2015 to 2016. Muranushi et al. [36] provided 24 h full-disk forecasting for M flare with 1 h cadence. The dataset consisted of 1574 positive samples and 6837 negative samples from 2011 to 2012.

Table 6.1 Forecasting precisions of solar flare for different forecasting period and flare level

Forecasting periods(hours)	Flare levels	Number of events				Performance index	
		TP	FN	TN	FP	TPR	FPR
6	C	1945	1106	37,516	7919	0.64	0.17
	M	304	60	39,782	8340	0.84	0.17
	X	29	4	40,894	7559	0.88	0.16
12	C	3386	1496	32,856	10,748	0.69	0.25
	M	559	115	38,388	9424	0.83	0.20
	X	61	6	41,050	7369	0.91	0.15
24	C	5338	2017	31,301	9830	0.73	0.24
	M	999	176	38,398	8913	0.85	0.19
	X	118	18	40,899	7451	0.87	0.15
48	C	7191	3591	31,607	6097	0.67	0.16
	M	1614	378	37,711	8783	0.81	0.19
	X	229	8	39,791	8458	0.97	0.18

Table 6.2 Performance comparisons of four 24 h M flare forecasting models

Reference	Definitions of events	Forecasting frequency	Forecasting periods	Testing samples	Evaluation index		
					TP rate	TN rate	TSS
[35]	Full disk	Every 6h	24h	2015–2016	80%	72%	0.525
	M class						
[36]	Full disk	Every 1h	24h	2011–2012	85%	67%	0.517
	M class			Cross validation			
[9]	Active region	24h before	24h	2010–2014	83.2%	93.3%	0.765
	M class			30% for testing			
Proposed model	Active region	Every 1.5h	24h	2010–2015	85%	81%	0.662
[1]	M class						

Bobra and Couvidat [9] established a forecasting model of 24 h M flare over active region. The dataset included 303 positive samples and 5000 negative samples, 70% data for training and remaining for testing. To validate our model, the dataset was split into ten folds, nine folds for training, one fold for testing. This process was repeated ten times, until each fold was tested once. Then, evaluation metrics were averaged over ten times of testing to give performance evaluation. The performance comparisons are listed in Table 6.2. It can be observed that the proposed model achieves the best prediction for flares, measured by TPR.

To better understand the mechanism of network, the first convolutional layer of the network was visualized in Fig. 6.2. The first convolutional layer consists of 64 convolutional kernels of size 11 × 11, which were initialized randomly as shown in Fig. 6.2a. Along with the increase of iterations, network weights became more and more orderly as shown in Fig. 6.2b,c. After 3000 iterations, the weights became

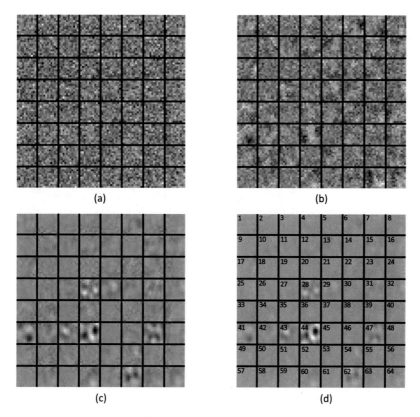

Fig. 6.2 Visualization of the proposed network (64 convolutional kernels of the first convolutional layer). (**a**) Initial random filter weights. (**b**) Filter weights after 1000 training iterations. (**c**) Filter weights after 2000 training iterations. (**d**) Filter weights after 3000 training iterations

stable gradually as shown in Fig. 6.2d, indicating that stable forecasting patterns were captured successfully.

6.1.4 Discussions

The proposed solar flare forecasting model has achieved state-of-the-art performance. It may be further improved by incorporating with more observations of solar chromosphere and corona. Thus, we further raised a new model over the first baseline model, by importing EUV images to cooperate with magnetograms for solar flare forecasting. Moreover, a dynamic deep network was raised for inferring solar flare from a magnetogram sequence. Since solar flare is a dynamic process, a sequence can better represent a dynamic process. For verifying this dynamic model, a database of solar flare sequences was established. The database contains more than 1091 flare sequences, covering all flares happening from 1996 to 2020. Each sequence has 30–100 magnetograms, and corresponding EUV images.

6.2 F10.7 cm Forecast

The source of F10.7 (wavelength equals to 10.7 cm) is gyroresonance emission and bremsstrahlung emission [37–39], which are two different mechanisms [40]. F10.7, as proxy of EUV observation, characterizes intensity of solar EUV radiation. It is also the input parameter of many existing ionosphere and navigation models. F10.7 forecast is of great importance to space weather. There has been a bunch of F10.7 forecast methods in the literature. Liu et al. [41] applied autoregressive model(AR) with 54-order to predict f10.7 within 27 days. Huang et al. [42] used support vector regression(SVM) method to predict F10.7 [42]. Lei et al. [43] defined an index PSR from 304 Å EUV observation to forecast F10.7 [43].

LSTM is a kind of RNN which is provided for processing time series data, exploring long-term dependencies of input sequence. It is now quite mature, used in pedestrian prediction [44], speech recognition [45], sequence to sequence learning [46]. In this work, it is employed to be the backbone, establishing a two-stream LSTM model for F10.7 forecast. The two streams have the same network structure, but the time steps of them are different, 3-days and 27-days, respectively. Compared to single stream model, two-stream model takes an additional 27-days input, which could mine knowledge about solar rotation cycle of 27-days, benefiting F10. 7 forecast. The experimental results demonstrate that two-stream model achieves better performance than single stream model.

6.2.1 Deep Learning Model for F10.7 cm Forecast

The proposed two-stream LSTM model for predicting F10.7 is shown in Fig. 6.3, which is composed of two streams. The number of nodes of the left stream and the right stream are different, depending on time steps. Each stream contains three layers, input layer, hidden layer and output layer. Input layer receives F10.7 sequence $(x_1, x_2, ..., x_T)$, $T = 3$ and $T = 27$ for two different streams. Hidden layer consists of LSTM cells, each of which gets one element of the input sequence. However, not all cells output, but only the last one. After hidden layer, we can get a compressed representation from input F10.7 sequence for each stream. These two representations are concatenated to be the input of output layer. In our task, output layer is a regressor, for predicting F10.7 of future 3 days.

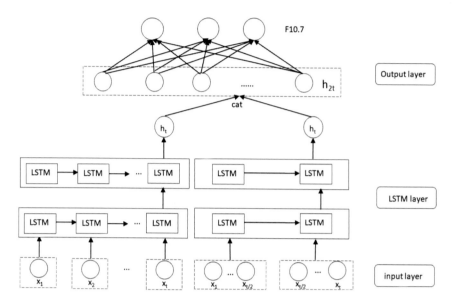

Fig. 6.3 Architecture of two-stream LSTM model

6.2.2   Model Validation

We implement the proposed model by using Python over PyTorch library for forecasting F10.7 of future 3 days. The database for model training and testing is obtained from SWPC (https://www.swpc.noaa.gov/). The daily F10.7 is collected from Jan. 1, 1996 to Apr. 30, 2018, including 8156 daily F10.7 observations. The database is split into training set and testing set. The training set is from Jan. 1,1996 to Dec. 30, 2014, while testing set is from Jan. 1, 2015 to Apr. 30, 2018. To measure model accuracy, mean square error (MSE) [47] between predicted F10.7 and ground-truth is employed. Adam optimizer is employed for training model, and batch-size is 128.

F10.7 has the cycle of 27 days. For exploring this cyclicity of F10.7, input data is a 54-dimension vector containing two cycles of daily F10.7 for both two streams. One has 54-dimension F10.7 input with time step of 1. The other has an input of 27 × 2 with time step of 2. For verifying advantage of the proposed two stream model, we compare it with the LSTM of single stream and traditional auto regression (AR) model. The codecs of these implementations can be accessed via (https://github.com/filterbank/F10.7Forecast). The two-stream LSTM model preforms the best, with the smallest MSE of 0.033870. The single stream LSTM with the MSE of 0.035045 is inferior to the two-stream one, but much better than AR model with the MSE of 0.097559. In addition, MSE tends to be stable after 50 epochs for the proposed model.

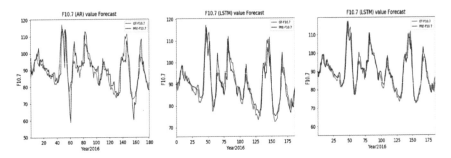

Fig. 6.4 F10.7 forecast by AR(left), LSTM(middle) and two-stream LSTM(right) (180-days daily F10.7 in 2016).

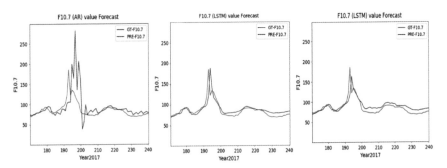

Fig. 6.5 F10.7 forecast by AR(left), LSTM(middle) and two-stream LSTM(right) (70-days daily F10.7 in 2017)

Figures 6.4 and 6.5 show 10.7 curves predicted by the proposed two-stream model and two benchmarks, where the left image gives the results of classical AR model, the middle one is for the single stream LSTM model, and the right one is for the proposed model. For comparison, real F10.7 in green color is also included for comparing with the predicted one in red color. It can be seen that the proposed model predicts F10.7 more close to ground truth than the AR model, thanking to exploration of long-term dependence of input sequence. There is a certain gap between predicted and real F10.7 for the AR model. As shown in Fig. 6.5, in June 2017, F10.7 had a sudden variation due to the sunspots and other interference factors. In this case, classical AR model fails, resulting in invalid prediction. Although, LSTM has a long-term dependence which can be calibrated to adapt to such a sudden variation, there is a certain delay of response for the two LSTM models as shown in Fig. 6.5. Comparing these two LSTM models, two-stream model is better than one-stream model, since it takes into account periodic characteristics of F10.7 additionally.

References

1. Huang X, Wang H, Xu L, Liu J, Li R, Dai X. 2018 Deep Learning Based Solar Flare Forecasting Model. I. Results for Line-of-sight Magnetograms. *Astrophys. J.* **856**, 7.
2. Giovanelli RG. 1939 The Relations Between Eruptions and Sunspots.. *Astrophys. J.* **89**, 555.
3. Gallagher PT, Long DM. 2011 Large-scale Bright Fronts in the Solar Corona: A Review of "EIT waves". *Space Sci. Rev.* **158**, 365–396.
4. Bloomfield DS, Higgins PA, McAteer RTJ, Gallagher PT. 2012 Toward Reliable Benchmarking of Solar Flare Forecasting Methods. *Astrophys. J. Lett.* **747**, L41.
5. Leka KD, Barnes G. 2003 Photospheric Magnetic Field Properties of Flaring versus Flare-quiet Active Regions. II. Discriminant Analysis. *Astrophys. J.* **595**, 1296–1306.
6. Leka KD, Barnes G. 2007 Photospheric Magnetic Field Properties of Flaring versus Flare-quiet Active Regions. IV. A Statistically Significant Sample. *Astrophys. J.* **656**, 1173–1186.
7. Mason JP, Hoeksema JT. 2010 Testing Automated Solar Flare Forecasting with 13 Years of Michelson Doppler Imager Magnetograms. *Astrophys. J.* **723**, 634–640.
8. Li R, Wang HN, He H, Cui YM, Zhan-LeDu. 2007 Support Vector Machine combined with K-Nearest Neighbors for Solar Flare Forecasting. *Chin. J. Astron. Astrophys.* **7**, 441–447.
9. Bobra MG, Couvidat S. 2015 Solar Flare Prediction Using SDO/HMI Vector Magnetic Field Data with a Machine-learning Algorithm. *Astrophys. J.* **798**, 135.
10. Nishizuka N, Sugiura K, Kubo Y, Den M, Watari S, Ishii M. 2017 Solar Flare Prediction Model with Three Machine-learning Algorithms using Ultraviolet Brightening and Vector Magnetograms. *Astrophys. J.* **835**, 156.
11. Qahwaji R, Colak T. 2007 Automatic Short-Term Solar Flare Prediction Using Machine Learning and Sunspot Associations. *Solar Phys.* **241**, 195–211.
12. Wang HN, Cui YM, Li R, Zhang LY, Han H. 2008 Solar flare forecasting model supported with artificial neural network techniques. *Advances in Space Research* **42**, 1464–1468.
13. Colak T, Qahwaji R. 2009 Automated Solar Activity Prediction: A hybrid computer platform using machine learning and solar imaging for automated prediction of solar flares. *Space Weather* **7**, S06001.
14. Ahmed OW, Qahwaji R, Colak T, Higgins PA, Gallagher PT, Bloomfield DS. 2013 Solar Flare Prediction Using Advanced Feature Extraction, Machine Learning, and Feature Selection. *Solar Phys.* **283**, 157–175.
15. Song H, Tan C, Jing J, Wang H, Yurchyshyn V, Abramenko V. 2009 Statistical Assessment of Photospheric Magnetic Features in Imminent Solar Flare Predictions. *Solar Phys.* **254**, 101–125.
16. Yu D, Huang X, Wang H, Cui Y, Hu Q, Zhou R. 2010 Short-term Solar Flare Level Prediction Using a Bayesian Network Approach. *Astrophys. J.* **710**, 869–877.
17. Guerra JA, Pulkkinen A, Uritsky VM. 2015 Ensemble forecasting of major solar flares: First results. *Space Weather* **13**, 626–642.
18. Falconer DA. 2001 A prospective method for predicting coronal mass ejections from vector magnetograms. *J. Geophys. Res.* **106**, 25185–25190.
19. Cui Y, Li R, Zhang L, He Y, Wang H. 2006 Correlation Between Solar Flare Productivity and Photospheric Magnetic Field Properties. 1. Maximum Horizontal Gradient, Length of Neutral Line, Number of Singular Points. *Solar Phys.* **237**, 45–59.
20. Huang X, Wang HN. 2013 Solar flare prediction using highly stressed longitudinal magnetic field parameters. *Research in Astronomy and Astrophysics* **13**, 351–358.
21. Huang X, Zhang L, Wang H, Li L. 2013 Improving the performance of solar flare prediction using active longitudes information. *Astron. Astrophys.* **549**, A127.
22. Raboonik A, Safari H, Alipour N, Wheatland MS. 2017 Prediction of Solar Flares Using Unique Signatures of Magnetic Field Images. *Astrophys. J.* **834**, 11.
23. Yu D, Huang X, Wang H, Cui Y. 2009 Short-Term Solar Flare Prediction Using a Sequential Supervised Learning Method. *Solar Phys.* **255**, 91–105.

24. Yu D, Huang X, Hu Q, Zhou R, Wang H, Cui Y. 2010 Short-term Solar Flare Prediction Using Multiresolution Predictors. *Astrophys. J.* **709**, 321–326.
25. Huang X, Yu D, Hu Q, Wang H, Cui Y. 2010 Short-Term Solar Flare Prediction Using Predictor Teams. *Solar Phys.* **263**, 175–184.
26. Korsós MB, Baranyi T, Ludmány A. 2014 Pre-flare Dynamics of Sunspot Groups. *Astrophys. J.* **789**, 107.
27. Korsós MB, Ludmány A, Erdélyi R, Baranyi T. 2015 On Flare Predictability Based on Sunspot Group Evolution. *Astrophys. J. Lett.* **802**, L21.
28. Barnes G, Leka KD. 2008 Evaluating the Performance of Solar Flare Forecasting Methods. *Astrophys. J. Lett.* **688**, L107.
29. Barnes G, Leka KD, Schrijver CJ, Colak T, Qahwaji R, Ashamari OW, Yuan Y, Zhang J, McAteer RTJ, Bloomfield DS, Higgins PA, Gallagher PT, Falconer DA, Georgoulis MK, Wheatland MS, Balch C, Dunn T, Wagner EL. 2016 A Comparison of Flare Forecasting Methods. I. Results from the "All-Clear" Workshop. *Astrophys. J.* **829**, 89.
30. Hinton GE, Salakhutdinov RR. 2006 Reducing the Dimensionality of Data with Neural Networks. *Science* **313**, 504–507.
31. LeCun Y, Bengio Y, Hinton G. 2015 Deep learning. *nature* **521**, 444.
32. Schmidhuber J. 2015 Deep learning in neural networks: An overview. *Neural Networks* **61**, 85–117.
33. He K, Zhang X, Ren S, Sun J. 2015 Spatial pyramid pooling in deep convolutional networks for visual recognition. *IEEE transactions on pattern analysis and machine intelligence* **37**, 1904–1916.
34. Jia Y, Shelhamer E, Donahue J, Karayev S, Long J, Girshick R, Guadarrama S, Darrell T. 2014 Caffe: Convolutional Architecture for Fast Feature Embedding. *arXiv e-prints* p. arXiv:1408.5093.
35. Murray SA, Bingham S, Sharpe M, Jackson DR. 2017 Flare forecasting at the Met Office Space Weather Operations Centre. *Space Weather* **15**, 577–588.
36. Muranushi T, Shibayama T, Muranushi YH, Isobe H, Nemoto S, Komazaki K, Shibata K. 2015 UFCORIN: A fully automated predictor of solar flares in GOES X-ray flux. *Space Weather* **13**, 778–796.
37. Schröter EH. 1971 On Magnetic Fields in Sunspots and Active Regions. **43**, 167.
38. Schunker H, Cally P. 2006 Magnetic field inclination and atmospheric oscillations above solar active regions. *Monthly Notices of the Royal Astronomical Society* **372**, 551–564.
39. Athay RG. 1976 The solar chromosphere and corona: Quiet sun. **53**.
40. Lee J. 2007 Radio Emissions from Solar Active Regions. *Space Science Reviews* **133**, 73–102.
41. Si-qing L, Qiu-zhen Z, Jing W, Xian-kang D. 2010 Modeling Research of the 27-day Forecast of 10.7cm Solar Radio Flux (I). *Chinese Astronomy and Astrophysics* **34**, 305–315.
42. Huang C, Liu DD, Wang JS. 2009 Forecast daily indices of solar activity, F10. 7, using support vector regression method. *Research in Astronomy and Astrophysics* **9**.
43. Lei L, Zhong Q, Wang J, Shi L, Liu S. 2019 The Mid-Term Forecast Method of F10.7 Based on Extreme Ultraviolet Images. *Advances in Astronomy* **2019**, 5604092.
44. Alahi A, Goel K, Ramanathan V, Robicquet A, Fei-Fei L, Savarese S. 2016 Social LSTM: Human Trajectory Prediction in Crowded Spaces. pp. 961–971.
45. Graves A, Jaitly N. 2014 Towards end-to-end speech recognition with recurrent neural networks. *31st International Conference on Machine Learning, ICML 2014* **5**, 1764–1772.
46. Sutskever I, Vinyals O, Le QV. 2014 Sequence to Sequence Learning with Neural Networks. *CoRR* **abs/1409.3215**.
47. Adhikari R, Agrawal R. 2013 *An Introductory Study on Time series Modeling and Forecasting.*

Printed in the United States
by Baker & Taylor Publisher Services